INTRODUCTION TO LINEAR ALGEBRA

Introduction to Linear Algebra

A PRIMER FOR SOCIAL SCIENTISTS

by

GORDON MILLS

Reader in Economics and Operational Research
University of Bristol

ALDINE PUBLISHING COMPANY
CHICAGO

George Allen and Unwin Ltd, London

First U.S. edition 1970
Aldine Publishing Company
529 South Wabash Avenue
Chicago, Illinois 60605

Library of Congress Catalog Card Number 75-105096
SBN 04 512008 0

Printed in Great Britain

PREFACE

Linear mathematical models play an important role in economics and in other social sciences. The aim of this book is to provide an introduction to those parts of linear algebra which are useful in such model-building, to illustrate some of the uses of the analysis, and to help the reader to learn for himself how to convert his formulation of a problem into algebraic terms. Prior knowledge of the social sciences is not necessary; the mathematical background which is required is mainly just school algebra; in particular, no use is made of calculus or of complex numbers. A novel feature of the mathematical content of the book is the treatment (in Chapter 7) of models expressed in terms of variables which must be whole numbers (integers). Throughout the book, there is emphasis on the computational aspects; to this end, determinants are not used in the main exposition, although they are treated in Appendix A.

An attempt has been made to bring in new ideas gently, often by way of example, before a general statement is given. At the same time, this is not a 'cook-book'; most theorems are proved, and an attempt has been made to combine rigour with simplicity of statement and a detailed step-by-step argument to help the beginner to find his way through the more complicated proofs. The reader should not simply *read* the book; instead he should keep pencil and paper at hand and test each piece of general argument by exploring not only the examples provided in the text but also others which he should construct for himself. Again, the exercises are not optional extras; it is necessary to practice by working through *all* of them. Furthermore, many of the exercises are used to develop points not included in the text, or to give practice in the algebraic formulation of applied problems. Appendix B gives solutions (or hints) for those exercises marked with asterisks.

On the whole, knowledge of any one chapter is required in the next chapter, and hence the material should be studied in the order in which it is presented. An important exception is the relatively abstract (and hence difficult) material in sections 2.5 to 2.8; this is not used again until section 4.6. The logical flow of the material in Chapter 7 is explained in the first section of that chapter.

The author is grateful for help received from a number of people. Several academic colleagues have read parts of the text; Professor John Black deserves special mention. Drafts of most of the chapters have been used by my students, who should be awarded much of the blame for any mistakes which remain. I am particularly grateful to all those who have helped with the physical production of the text, and especially to Mrs Anne Kempson who has produced excellent typescript of difficult material. She has also pointed out that the baker's recipes (given in section 1.8) are unlikely to lead to palatable results; readers are hereby warned.

G.M.
Bristol
September 1968

CONTENTS

CHAPTER 1

Introduction

1.1 *Linear and non-linear systems*

The preface to this book contains some important remarks on how to use the book; if you have not already studied these remarks, you should do so now. The book deals with the algebra of *linear* systems; it also illustrates the use of these concepts and techniques in economics and in other social sciences. The algebra to be studied is called 'linear' because graphs and sketches giving geometrical interpretation or illustration are composed of straight lines. In algebraic language, attention is confined to expressions in which each term is linear, i.e. each variable is raised to the power one, and any one term in the expression contains only one variable (unless it happens to be a constant term which, of course, does not contain any variable at all). The following example of a linear expression has four terms and involves three variables:

$$3+2x_1+x_2+4x_3$$

Thus the book is *not* concerned with expressions having non-linear terms such as

$$\sqrt{x_3},\ {x_1}^2,\ {x_1}^{-1},\ 7x_1x_2,\ \log x_1 \text{ or } 2x_2e^{x_3}$$

The mathematical analysis of linear expressions is old-established. In the great bulk of it, it is supposed that the variables x_1, x_2, x_3 etc. are *continuous*; that is to say, each variable may take on *any* value, including not only whole numbers (or integers, as they are often called) such as 2, 23, 175 and so forth, but also fractions such as $\frac{1}{2}$ and decimal expressions such as $2{\cdot}35\dot{3}$. In intuitive language, a variable may take on any value within a specified range or interval. It is usually supposed that the relevant range is all positive or negative numbers, however large or small. Similarly, in this book, except where otherwise indicated, the mathematical analysis is developed on the assumption that we are dealing with continuous variables which can take on positive or negative values. Of course, in applying the analysis to social science contexts, negative values for variables often have to be ruled out as inadmissible; there is no sense in predicting that next year's rainfall will be −37·3 cm. or that national income will be −£247·80 per head. Thus in considering the solution of a set of linear equations (for example) the applied context may make it necessary to ask not merely 'Is there a solution?' but rather 'Is there a non-negative solution?' On some occasions, the context may require the variables to take on integer values only. Often the usual mathematical analysis in terms of continuous variables may give an adequate approximation; at the end of the calculation, any non-integer values can be rounded to the nearest integer. But sometimes it is necessary to recognize the integer requirement (i.e. to treat the variables as *discrete* rather than continuous variables) right at the outset, in which case a very different type of mathematical analysis is required. On the whole, the analysis of linear systems involving discrete variables is not as well developed as that for continuous variables; but a short introduction to the subject is included in Chapter 7.

Linear systems are very useful in analysing many problems in economics and other social sciences, and in carrying out statistical estimation. In other words, many phenomena can be represented (at least to a good approximation) by linear functions. This is very fortunate because it turns out that both theoretical analysis and numerical calculation are much simpler for linear than for non-linear models. But in any given problem context, we must not automatically *assume* that a linear model will be sufficient to give an adequate representation of the phenomena being studied. Instead, as the model is constructed, we must ask at each stage whether it is appropriate to use a linear representation; some of the kinds of question which arise are illustrated by the example in the next section. And when a linear model cannot provide even an approximate representation, then we must formulate the problem in non-linear terms, and face up to the mathematical and computational difficulties which ensue.

1.2 *Simultaneous linear equations: an economic example*

Elementary algebra courses introduce small examples of the solution of simultaneous linear equations: for example, find the values of x_1 and x_2 which satisfy

$$2x_1 + 3x_2 = 4$$
$$2x_1 + x_2 = 2$$

By subtracting the second equation from the first, we solve for x_2 to find $x_2 = 1$, and hence, by substituting for x_2, we find $x_1 = \frac{1}{2}$ to complete the solution.

TABLE 1.1

	Purchases by each industry per unit of its own output: Industry 1	Industry 2	Quantities which final consumers wish to purchase
Product of:			
Industry 1	0·4	0·8	50
Industry 2	0·2	0·2	90

Many applied problems turn out to require the solution of such sets of simultaneous linear equations. Consider now an economic example. Suppose an economy regarded as two industries each of which produces commodities for final consumption. In order to be able to produce, each industry consumes some of the product of the other industry (e.g. the steel industry uses coal); and also it consumes some of its own product (e.g. a power station which generates electricity also uses some electricity whilst doing so). In particular suppose this industrial consumption is strictly proportionate to output and that the particular proportions are as shown in Table 1.1; in other words for each (physical) unit of its own output, industry 1 requires 0·4 units of its own output and 0·2 units of the output of industry 2, as shown in the first column of the table. The second column shows the proportions for industry 2, while in the third column, the required final output for the use of consumers is shown. Of course, each industry also uses 'non-produced' resources such as labour, but this is ignored here in the interests of simplicity. With this technology, the problem is: how much should each industry produce gross (i.e. in total) in order to provide these quantities for consumers? Clearly each industry will have to produce more than the quantities demanded by consumers in order to allow for the amounts used up in production. The total production can be calculated by solving a set of simultaneous linear equations. Suppose that gross production of industry 1 is x_1 units, and that of industry 2 is x_2 (of its units). Use of the production of industry 1 amounts to final consumption + amount used in industry 1 + amount used in industry 2

$$= 50 + 0{\cdot}4\,x_1 + 0{\cdot}8\,x_2$$

This is equal to gross production (if we assume no more is produced than is required) and hence we have our first equation:

$$x_1 = 50 + 0{\cdot}4\,x_1 + 0{\cdot}8\,x_2$$

The second equation is developed by similarly considering the production of industry 2:

$$x_2 = 90 + 0{\cdot}2\,x_1 + 0{\cdot}2\,x_2$$

These two equations may be solved (by the elementary method outlined at the beginning of this section) to show that the (gross) production required of industry 1 is $x_1 = 350$ units, and that for industry 2 production is $x_2 = 200$ units. (Check this for yourself.)

In building up the model in the way described above, we have supposed that a *linear* model gives an adequate representation of the situation of these two interacting industries. In particular, we have supposed that two properties possessed by all linear models, namely *linear homogeneity* and *additivity*, are appropriate to the situation. In the present mathematical context, *linear homogeneity* means that if a variable at a prescribed value x_1 has a certain effect, and if the variable is changed to an amount αx_1, then the amount of the effect is also multiplied by α (where α is any real number). In the present case, the homogeneity assumption enters into the relationship between the input and output quantities of an industry: we have supposed that if (for example) industry 1 doubles its output level, it will require *exactly* twice as much of the inputs. Whether this is a fair assumption is a matter for empirical investigation in the industry; if it turned out to require (say) less than twice as much, then the homogeneity assumption is not valid and we could not build a *linear* model (except perhaps as an approximation to the reality). The *additivity* property implies that if a variable x_1, at a certain value and taken by itself, gives a certain effect and if similarly a certain value for x_2 also by itself has some specified effect, then if these two variables are taken together at the prescribed values, the total effect is simply the sum of the two separate effects. Again, this can be clarified and made more precise in terms of the present example: if the first industry produces x_1 units of its output, it requires as one of its inputs, $0{\cdot}2\ x_1$ units of the second industry's output; if, separately, the second industry produces x_2 units of its output, it requires $0{\cdot}2\ x_2$ units of its own output (and in these relations we employ the homogeneity assumption). Now suppose the two industries are both hard at work, producing x_1 and x_2 units respectively; we go on to make the additivity assumption, that the total industrial demand for the product of industry 2 is simply the sum of the separate requirements, namely $0{\cdot}2\ x_1 + 0{\cdot}2\ x_2$. Putting it another way, the supposition is that industry 2's requirement is not affected by the level x_1 at which industry 1 is working (and vice-versa); mathematically speaking it is assumed that in the expression for the total requirement, there is no term such as $x_1 x_2$ or any other term which expresses such an interaction effect between the two industries. In the present case, the additivity assumption seems very plausible. Indeed this assumption is often appropriate even when the homogeneity assumption is not a proper description of the situation to be described. Nevertheless, additivity is not universally present and we must always stop to consider whether the assumption seems justified.

Finally, note that the input coefficient for each industry's use of its own product (for example, 0·4 input units of the product of industry 1 per unit output of that industry) is less than unity in each case. In common-sense terms, if it took more than a ton of coal (say) to produce a ton of coal, there could be no (*positive*) output from that industry. And clearly this is one of those cases where negative values for the variables make no practical sense. With the coefficients as stated, all is well: positive values are obtained for x_1 and x_2. But check for yourself to see what happens in the solution of the equations if this coefficient 0·4 for industry 1 becomes a coefficient of (say) 1·4.

1.3 *Simultaneous linear equations: general systems*

The example of the previous section is a small, primitive case of what is called input–output analysis. In more realistic models there may be several hundred industrial and other sectors, leading to a system of hundreds of equations involving hundreds of variables.

For such large systems, it is convenient to have a compact notation. First let us introduce the idea of 'double-subscripted' coefficients; for the above example it is convenient to define a set of coefficients as follows:

a_{ij} represents the number of units of the product of industry i, purchased by industry j for each unit of output of industry j.

For the data in Table 1.1, $a_{12} = 0{\cdot}8$. Where $j = i$ we are concerned with an industry's requirement for its own product; thus $a_{11} = 0{\cdot}4$. Notice that the first subscript i refers to the row in the table, and the second subscript j refers to the column; this convention is almost universally used.

Continuing with the example, let b_j denote the number of units of final consumer demand for the product of industry j, and suppose that there are a total of n industries. The model now comprises a set of n simultaneous linear equations. The first equation relates to the output of industry 1, and says that the total output equals the sum of the amounts used up by the n industries plus the amount consumed:

$$x_1 = a_{11}x_1 + a_{12}x_2 + a_{13}x_3 + \ldots + a_{1n}x_n + b_1$$

Each of the other $n-1$ equations can be written down in similar fashion:

$$x_2 = a_{21}x_1 + a_{22}x_2 + a_{23}x_3 + \ldots + a_{2n}x_n + b_2$$

$$\vdots$$

$$x_n = a_{n1}x_1 + a_{n2}x_2 + a_{n3}x_3 + \ldots + a_{nn}x_n + b_n$$

(Check your understanding of this by writing out the third equation for yourself.) For a large system such as this, it becomes tiresome to write it out in full. A shorthand notation introduced in the next section gives a much more compact representation.

1.4 *Summation notation*

Suppose we come across the expression

$$c_1 + c_2 + c_3 + c_4 + c_5 + c_6 + c_7 + c_8$$

Since it is tedious to write it out in full, a common device is to show just a few terms to indicate the nature of the sequence:

$$c_1+c_2+\ldots.+c_8$$

An even more compact (and more precise) notation is to use the *summation sign* Σ (the capital Greek letter 'sigma'). The expression may then be written $\sum_{i=1}^{8} c_i$ where $\sum_{i=1}^{8}$ means 'add up over all values of i from 1 to 8'. The letter i is used here to represent the subscripts; in other words the typical (i^{th}) term in the expression is regarded as c_i. But any other letter would serve just as well, provided we are consistent in the use of notation.

Next consider the expression:

$$2c_1+2c_2+2c_3+2c_4$$

Clearly this may be written as $\sum_{j=1}^{4} 2c_j$ (where this time j has been chosen to represent the subscripts, just for a change). Alternatively, however, this may be written $2\sum_{j=1}^{4} c_j$ since the coefficient 2 is constant in each term in the expression and hence can be brought outside (i.e. in front of) the Σ; in this respect the Σ here acts in the same way as a bracket does in arithmetic or elementary algebra. On the other hand, the expression $a_1c_1+a_2c_2+a_3c_3$ can be written $\sum_{i=1}^{3} a_ic_i$ but here the a_i varies with i and hence is not constant with respect to the summation; thus it must follow the Σ where it is included as part of the summation. Occasionally, quite clever tricks can be played with this summation notation; for example

$$c_1+2c_2+3c_3+4c_4 = \sum_{j=1}^{4} jc_j$$

Now let us return to the general system of simultaneous linear equations written out at the end of the previous section. The first equation of the system can be written

$$x_1 = \sum_{j=1}^{n} a_{1j}x_j+b_1$$

(Note that the coefficient a_{1j} varies from one term to the next as j

varies; in other words it is not constant with respect to the summation, and so it must remain after the summation sign.) The typical equation of the system can be regarded as the ith equation and written

$$x_i = \sum_{j=1}^{n} a_{ij}x_j + b_i$$

Thus the entire system of n equations can now be very briefly described by the statement

$$x_i = \sum_{j=1}^{n} a_{ij}x_j + b_i \qquad (i = 1, 2, \ldots, n)$$

where the annotation in brackets indicates the range of values for i, each of which defines a different equation in the system. To sum up, *within* one equation, the j subscripts vary and the summation is with respect to j; *between* equations, the i subscript varies, or in other words, each different value of i defines a different equation. Compare this brief description of the system of equations with the full statement of it at the end of the previous section.

1.5 *Exercises*

1. Write out the following summation expressions in full:

(a) $\sum_{i=2}^{4} x_i$ (b) $\sum_{j=8}^{10} a_j x_j$ *(c) $\sum_{i=0}^{2} x^i$

*(d) $\sum_{i=1}^{3} (x_i + i)$ (e) $\sum_{k=0}^{4} by_k$

(Note that the expression in case (c) is *not* linear in the variable x.)

2. Write out the following expressions using Σ notation:

*(a) $a_{0,1}x_1 + a_{0,2}x_2 + a_{0,3}x_3 + a_{0,4}x_4$
(b) $x^{-1} + x^{-2} + x^{-3} + x^{-4}$
(c) $a_{1,2}y_{2,2} + a_{1,3}y_{2,3} + a_{1,4}y_{2,4} + a_{1,5}y_{2,5}$

3. By writing out the full expressions in each case, show that the following equations are true:

(a) $\left(\sum_{i=0}^{n-1} y_i\right) + y_n = \sum_{i=0}^{n} y_i$

(Note that the brackets on the left-hand side are not strictly necessary; they are to be understood even when not placed there. If the summation sign is to apply to an expression of more than one term, then brackets *must* be included, as in the next expression.)

(b) $\sum_{i=1}^{3} (x_i + y_i) = \sum_{i=1}^{3} x_i + \sum_{i=1}^{3} y_i$

*4. Suppose that a commodity is sold on a market at price p and that the quantity offered for sale q_s is given by the linear equation

$$q_s = -a + bp$$

where it is supposed that a and b are both strictly positive, i.e. greater than zero. Also suppose that the quantity demanded q_d behaves in just the opposite way, that is it is smaller the larger is the price on the market, and that again it is given by a linear equation

$$q_d = c - dp$$

where c and d are both strictly positive. Note that these linear expressions for q_d and q_s are valid only in the range of p which gives positive values for q_d and q_s. And in particular note that those who offer the commodity are not willing to supply anything at all until the price is high enough to make bp greater than a, in other words to make q_s positive. (Consider on a graph where these lines must lie relative to the origin.) The price on the market (it is supposed) adjusts until equilibrium is reached, i.e. until $q_d = q_s$ so that everyone who wants to sell at that price is able to find a buyer for the appropriate quantity, and vice-versa. Solve the system of three linear equations in order to find the price which gives this equilibrium situation, and carefully justify each step in your manipulation of the equations. Clearly, the situation requires that the price be positive (or at any rate non-negative). Does this linear model guarantee that the solution will meet this requirement? If so, why?

*5. Consider a simple input–output model (of the type discussed in sections 1.2 and 1.3) in which the economy comprises three industries having a_{ij} (input–output coefficients) as shown in the table. The quantities required by the final consumers of the products of industries 1, 2 and 3 are 70, 20 and 90 units respectively. What are the required gross production levels for each industry?

	Coefficients a_{ij} for purchases by		
	Industry 1	Industry 2	Industry 3
Product of:			
Industry 1	0·2	0·6	0·2
Industry 2	0	0·2	0·4
Industry 3	0·2	0·2	0

1.6 *Inequalities*

Elementary algebra is often concerned with equations. But *inequalities* are also important, especially in applications in economics and related subjects. The following signs are used:

$>$ is greater than
$<$ is less than
$\geqq$ is greater than or equal to
$\leqq$ is less than or equal to

For example, if we wish to remark that '5 is greater than 3' it may be written

$$5 > 3$$

and the other signs are used in a similar way. The first two are sometimes described as *strict inequalities*. The latter pair are known as *weak inequalities* since the case of equality is permitted; for example

$$x \geqq 2$$

means that x has either the value 2 (the equality case) or any value greater than 2.

As with the algebra of equations, that for inequalities has certain properties and rules which have to be observed:

A: Inequalities are *transitive*. For example, if $a > b$ and $b > 2$, then $a > 2$. On the other hand, if $a < b$ and $b > 2$, it is not possible to say whether or not $a > 2$. In other words, the direction of the inequalities has to be appropriate before we can exploit the transitivity property.

B: Addition and subtraction. If $a > b$, then $a+e > b+e$, and $a-e > b-e$. Further, two or more inequalities may be added provided that the inequalities are arranged to have the same direction; for example, if $a > b$ and $c > d$, then $a+c > b+d$. Similar rules hold for weak inequalities.

C: Multiplication and division. If an inequality is multiplied by a positive number, then the direction of the inequality is unchanged. For example, consider $5 > 3$; if this is multiplied by 2, the result is $10 > 6$ according to the rule; clearly this is valid. But if the multiplier is negative, the direction of the inequality is reversed; for example, multiply $5 > 3$ by -2 to give $-10 < -6$. (If you have any difficulty here, think of these numbers marked out on the scale of a graph; this immediately demonstrates the validity of the result.) Note that an inequality may be multiplied by -1 in order to change the direction of the inequality sign; this trick can be useful if we wish to put the given inequality into the opposite form, in order to match it up with other given inequalities. Sometimes, we multiply by a variable and if the sign of the variable is not known, then particular care must be taken to specify the possible cases:

$$\text{if } a > b, \text{ then} \quad \begin{array}{ll} ka > kb & \text{if } k > 0 \\ ka = kb & \text{if } k = 0 \\ ka < kb & \text{if } k < 0 \end{array}$$

Division by k can be treated as multiplication by k^{-1}. Similar rules apply to weak inequalities.

D: Squaring. If $a > b$ and $a > 0$, $b \geqq 0$, then $a^2 > b^2$. But if $a < 0$ and $b < 0$, then $a^2 < b^2$; for an example, consider $a = -3$ and $b = -4$. Various other cases are possible; also the rules may be extended to the multiplication of one inequality into another inequality.

E: Solving an inequality. The rules developed above may be applied. For example,

if $4x-3 > x+3$
add 3 to both sides: $4x > x+6$
subtract x from both sides: $3x > 6$
divide both sides by 3: $x > 2$

thus giving the solution, which is of course a *range* of values for x.

Examples:

(1) If $u > 4$, show that $2u + 7 > 10$.

Since $u > 4$,	$2u > 8$	(multiplying by $+2$)
Thus	$2u + 7 > 15$	
But	$15 > 10$	
$\therefore$	$2u + 7 > 10$	

(2) If $0 < x < 1$ (in other words, if x lies between 0 and 1), show that $x^2 < x$. We are given that

$$x < 1$$

Now multiply both sides by x (remembering that $x > 0$ and hence the direction of the inequality remains the same), to give

$$x^2 < x$$

1.7 *Linear inequalities: graphical representation*

Inequalities are met in both linear and non-linear algebra. *Linear* inequalities (i.e. involving linear terms only) occur frequently, especially in applications in social sciences contexts. For such a linear inequality involving n variables x_i where $i = 1, 2, \ldots, n$, the general form may be denoted

$$a_1x_1 + a_2x_2 + a_3x_3 + \ldots + a_nx_n \leqq b$$

This may be written in summation notation as

$$\sum_{i=1}^{n} a_ix_i \leqq b$$

When the inequality involves only two variables, it is sometimes helpful to represent it graphically. As an example, consider

$$2x_1 + x_2 \leqq 4$$

The first step is to plot the straight line which corresponds to the *equality*. To do this, find *any* two points on the line: if $x_1 = 0$, then $x_2 = 4$; and if $x_2 = 0$, then $x_1 = 2$. Thus the line must pass through these two points, and this enables us to draw it on the graph as in Figure 1.1. The next problem is to decide which points on the graph

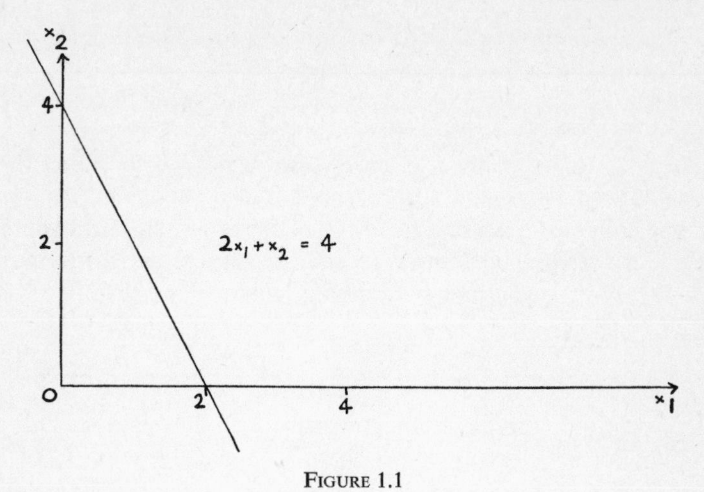

FIGURE 1.1

satisfy the *inequality*. Clearly the origin $(x_1, x_2) = (0, 0)$ is one such point since the left-hand side evaluated at the origin is

$$(2 \times 0) + 0 \leqq 4$$

Similarly, we find that all points which lie on or below the line satisfy the inequality. Conversely, if the inequality is

$$2x_1 + x_2 \geqq 4$$

then the origin (for example) does not satisfy the inequality; this time it turns out that all points on or above the line satisfy the inequality. In either case points with one or more negative coordinates may satisfy the inequalities; for example the point $(-2, 4)$ satisfies the inequality $2x_1 + x_2 \leqq 4$. If the inequality were

$$2x_1 + x_2 < 4$$

points *on* the line itself would not satisfy the inequality.

In summary, a linear inequality in two variables divides two-dimensional space into two regions, one of which satisfies the inequality and may be called the feasible region. Similar remarks apply in three-dimensional space, except that the related equality corresponds to a plane rather than to a straight line. In n-dimensional space, we speak of a hyper-plane as representing geometrically the linear equality; the plane and the line in three- and two-dimensional space are then just special cases of hyper-planes.

1.8 *Linear inequalities: an economic example*

A baker has 15 lb. of flour, 18 oz. of sugar and 20 oz. of butter. He is unable to purchase any further supplies of these ingredients, but can buy unlimited quantities of any other ingredients he may require. He has two recipes, one for a cake and one for a scone, the proportions of flour, sugar and butter being summarized in Table 1.2.

TABLE 1.2

	Weight of ingredients required per dozen		Weight of available supplies of ingredients
	Scones	Cakes	
Flour (lb.)	0·5	0·3	15
Sugar (oz.)	0·5	1·0	18
Butter (oz.)	0·4	1·25	20

He realizes that the more cakes he makes the fewer the scones he can make, and vice-versa. More generally, he wonders precisely what options are open to him. This can be explored by formulating some linear inequalities and recording them on a graph. Let

$$x_1 = \text{no. of dozens of scones made}$$
$$x_2 = \text{no. of dozens of cakes made.}$$

Consider first the flour requirements:

amount used in scones + amount used in cakes $\leqq$ amount available (lb.)

$$0{\cdot}5\,x_1 + 0{\cdot}3\,x_2 \leqq 15$$

Thus as far as the flour limit is concerned any values for x_1 and x_2 are feasible provided that they satisfy this inequality. But there are also limits on sugar and butter. Similar arguments lead to the inequalities

$$0{\cdot}5\,x_1 + \quad x_2 \leqq 18$$
$$0{\cdot}4\,x_1 + 1{\cdot}25\,x_2 \leqq 20$$

Also the baker cannot make a negative number of either scones or cakes. In other words, a negative value for x_1 or x_2 would have no physical meaning, and must therefore be ruled out in the formal statement of the problem. Thus we require two further inequalities (sometimes called the non-negativity requirements):

$$x_1 \geqq 0$$
$$x_2 \geqq 0$$

These five inequalities can now be recorded on a graph in which x_1 and x_2 are recorded along the axes. Prepare such a graph for yourself as an exercise and then check it against Figure 1.2. The feasible region

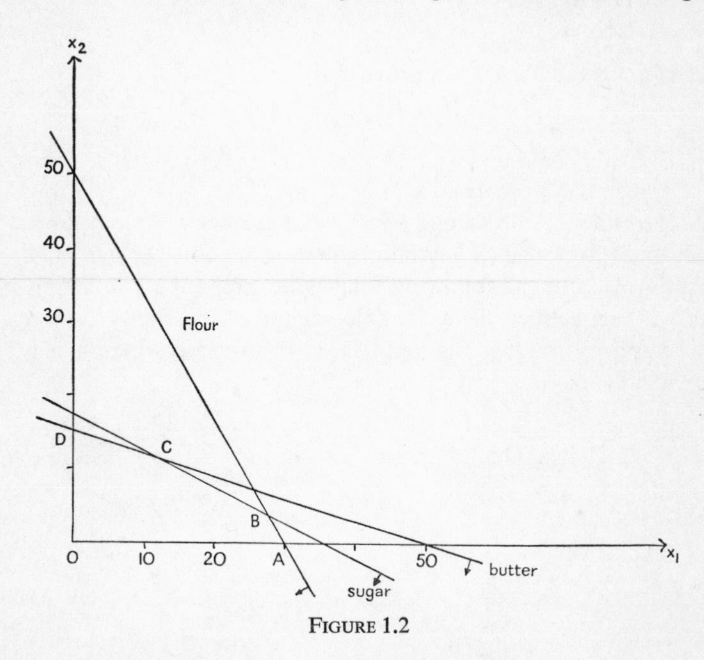

FIGURE 1.2

(i.e. the region of all values for x_1 and x_2 which meet all the restrictions) is marked in the Figure as the area OABCD. (You should check in your graph that this area is on the 'correct' side of each of the five lines representing the related equalities.)

This result shows that the baker could make 30 dozen scones and no cakes (point A), or 16 dozen cakes and no scones (point D) or any other combination falling in the feasible region. Note that it is impossible for the baker simultaneously to use up all his supplies of flour, sugar and butter. At point B, for example, he uses up all his flour and sugar but has some butter to spare – this is shown on the graph by B being *on* the flour and sugar constraint lines (as they are sometimes called) but inside the region corresponding to the butter inequality.

1.9 *Exercises*

1. Solve the inequalities

 (a) $2x+7<5x+1$
 *(b) $(x-3)^2>x^2-9$
 *(c) $x^2 \geqq 4$

2. If $u>0$ and if $u+7>v$, prove that

$$u(u+1)>u(v-10)$$

*3. If $1-u<7-v$ and $4v+13>(w+2)^2-w^2+1$, what is the relationship between u and w?

*4. If $a>3u$, $7-b<4u$ and $u<2$, what can be said about the value of $(a+b)$? What (if anything) can be said about the sign of b?

5. For each of the following four cases, prepare a graph showing the inequalities. If a feasible region exists, show it on your graph by shading the area. Discuss any case where a feasible region does not exist.

(a)
$$\begin{aligned} 2x_1+x_2 &\leqq 10 \\ x_1/2+2x_2/3 &\leqq 4 \\ x_1, x_2 &\geqq 0 \end{aligned}$$

(b)
$$\begin{aligned} 2x_1+3x_2 &\geqq 12 \\ x_1 &\leqq 5 \\ x_2 &\leqq 4 \\ x_1, x_2 &\geqq 0 \end{aligned}$$

*(c)
$$\begin{aligned} x_1+x_2 &\leqq 8 \\ x_1-x_2 &\geqq 0 \\ x_1 &\leqq 7 \\ x_2 &\geqq 6 \end{aligned}$$

*(d)
$$\begin{aligned} x_1-x_2 &\geqq -1 \\ -2x_1/3+x_2 &\leqq 2 \\ x_1, x_2 &\geqq 0 \end{aligned}$$

6. A pottery manufacturer makes two types of plate. The first type is plain and requires 2 units of labour per 1000 plates per day; the plant capacity for the manufacture of these plates is restricted to a maximum of 2000 plates per day. The second type is made by taking finished plates of the first type and adding some decorations by hand; the task of decoration requires 3 units of labour per 1000 plates decorated per day. Both types of labour are

provided by the same labour force which yields 6 units of labour per day. Give an algebraic formulation of these production circumstances and draw a graph which illustrates all possible production plans.

*7. Consider the two-industry input–output model specified in section 1.2, and draw a graph to represent the two *equations* which describe the production balance of each industry. Hence find the region of feasible gross production levels which obtains if (instead of balance) we require that, for each industry,

gross production $\geqq$ amounts used in industry + final consumption

For the general formulation of the two-industry model (i.e. having final consumption quantities b_1 and b_2, and input–output coefficients a_{ij}, as defined in section 1.3), what condition must be met to ensure that such a feasible production region exists?

1.10 *Some concepts from set theory*

Although the theory of sets is often of importance in more advanced work in linear algebra (and in other parts of mathematics), this book does not use very many concepts from such theory. However it is helpful to introduce in this section just a few of the most basic ideas.

The idea of a set is so fundamental that it is rather difficult to explain. For a formal definition it is best perhaps simply to say:

Definition. A *set* is a collection of distinct objects.

The objects or *elements* of the set may be listed or enumerated as a way of describing the set. For example, a set might be the integers 2, 3, 4 and 5. We might let this set be called S and write it $S = \{2, 3, 4, 5\}$; it is a common convention to use a capital letter to denote the set, and to use curly brackets to enclose the list of the elements. Note that the elements must be *distinct*; the list of integers 2, 3, 4, 3 is *not* a set. Another notational convention which can sometimes be employed is to describe the set by quoting just a typical element. For example a set to be denoted A might comprise n elements to be denoted a_i where $i = 1, 2, \ldots, n$. The set might then be written $A = \{a_i\}$, although by itself this does not indicate how many elements there are. The kinds of objects or elements which might be regarded as making up a set are extremely diverse. Examples of sets are: the people on board an aircraft; all non-negative integers not greater than 100; the countries which are members of the United Nations; the houses in a road; the players in a football team; the volumes of an encyclopaedia; the five inequalities appearing in the example in section 1.8. Membership of a set may be indicated by the symbol $\in$ which means 'is an element of'. If the set comprising the members of a football team is denoted by V, and if John is playing in the team, we may write John $\in V$.

There are some important basic relationships between sets:

Definition. The sets A and B are said to be *equal*, written $A = B$, if they comprise the same elements, i.e. if every element of A is also an element of B, and vice-versa.

Note that the order in which the elements are listed is immaterial. For example, if $S = \{2, 3, 4, 5\}$ and if $T = \{2, 4, 3, 5\}$ then $S = T$.

Definition. A *subset* B, of a set A, is a set all of whose elements are in A, while not all the elements of A are necessarily in the subset B.

For example, if John leaves the field because he is injured, the remaining football players are a subset of V. Again if $S = \{2, 3, 4, 5\}$ and if $W = \{2, 5\}$, then W is a subset of S. In fact W is what is often called a *proper* subset, to emphasize that it does not contain all the elements of S. This distinction is made because the definition of a subset permits the whole of the given set A to be regarded as a subset of A. In practice, however, almost all the subsets we might wish to identify are proper subsets. Two further pieces of notation are useful here; they are the set inclusion symbols:

$\subset$ 'is contained in'
$\supset$ 'contains' or 'includes'

If T is a subset of S, we may write $T \subset S$ or $S \supset T$. (A useful mnemonic for distinguishing between these two symbols is to associate $\subset$ with $<$, and to associate $\supset$ with $>$. If T is a proper subset of S then $T \subset S$ can be associated with the idea that the number of elements in T is less than the number in S. But remember of course that the two relationships denoted by $<$ and by $\subset$ are mathematically quite separate and distinct, and must never be confused with each other.)

Definition. If two sets have no elements in common, they are said to be *disjoint* sets.

For example, if $S = \{2, 3, 4, 5\}$ and $T = \{6, 7\}$, then S and T are disjoint sets. Note that these three definitions are of relations *between* sets; they do not describe characteristics which a set may have absolutely, but merely compare the set with another prescribed set. For example, with S and T as just defined, T is a member of a pair of disjoint sets. But if $W = \{6, 7, 8\}$ then T is a subset of W. The character of T depends on the other set or sets with which it is compared.

Certain operations on sets can now be introduced.

Definition. Given two sets A and B, the *union* of the sets, written $A \cup B$ (and read 'A union B'), is the set comprising all elements in A or B or both.

Examples:

(1) If $A = \{2, 3, 4, 5\}$ and $B = \{4, 5, 6\}$, then $A \cup B = \{2, 3, 4, 5, 6\}$
(2) If $C = \{1, 7, 13\}$ and A as before, $A \cup C = \{2, 3, 4, 5, 1, 7, 13\}$
(3) If $D = \{2, 3\}$, $A \cup D = \{2, 3, 4, 5\} = A$, since D is a subset of A

Definition. Given two sets A and B, the *intersection* of the sets, written $A \cap B$ (and read 'A intersection B') is the set comprising all those elements which are in both A and B.

Examples:

(4) $A \cap B = \{4, 5\}$ since these are the only elements present in both A and B
(5) $A \cap D = \{2, 3\}$, which equals D since D is a subset of A
(6) A and C have no common elements i.e. they are disjoint sets. Thus the intersection has no elements. It is useful to have a name for such a set; the usual term is a *null set*, denoted $\varnothing$. Thus we write $A \cap C = \varnothing$.

(A useful mnemonic for distinguishing between $\cup$ and $\cap$ is to associate $\cup$ with the letter U which is the first letter in the word union.) These two set operations have the same commutative, associative and distributive properties as are found in arithmetic and elementary algebra. Specifically, the commutative property means that the order in which the entities are listed is immaterial; in elementary algebra $a+b = b+a$ and $a \times b = b \times a$. Similarly for these set operations

$$A \cup B = B \cup A \quad \text{and} \quad A \cap B = B \cap A$$

as may be seen immediately from the definitions of the operations. The associative property means that when more than one application of an operation is made, the sequence of applications is of no consequence. Thus in elementary algebra, $a+(b+c)=(a+b)+c$ and $a\times(b\times c)=(a\times b)\times c$. For the two set operations:

$$A\cup(B\cup C)=(A\cup B)\cup C$$
$$A\cap(B\cap C)=(A\cap B)\cap C$$

In other words, it is not necessary to include the brackets at all; these results again follow directly from the definitions. Finally, the distributive property is illustrated by $a\times(b+c)=(a\times b)+(a\times c)$. For the set operations we have

$$A\cup(B\cap C)=(A\cup B)\cap(A\cup C)$$

and

$$A\cap(B\cup C)=(A\cap B)\cup(A\cap C)$$

Again these follow from the definitions, though a little less immediately. The following examples illustrate.

Examples:

(7) With the sets defined as before, $A\cup B=\{2, 3, 4, 5, 6\}=B\cup A$

(8) $A\cup(B\cup C)=A\cup\{4, 5, 6, 1, 7, 13\}=\{2, 3, 4, 5, 6, 1, 7, 13\}$
$(A\cup B)\cup C=\{2, 3, 4, 5, 6\}\cup C=\{2, 3, 4, 5, 6, 1, 7, 13\}$

(9) $A\cap(B\cup C)=A\cap\{4, 5, 6, 1, 7, 13\}=\{4, 5\}$
$(A\cap B)\cup(A\cap C)=\{4, 5\}\cup\varnothing=\{4, 5\}$

1.11 *Necessary and sufficient conditions*

The concepts of 'a necessary condition' and 'a sufficient condition' are used extensively in logical arguments, and the theorems given in the following chapters of this book are no exception. Thus it is useful to clarify the concepts at this stage. In plain language, a necessary condition is something which must hold for a result to follow, but it may not guarantee the result. On the other hand, a sufficient condition guarantees the result, but the result may perhaps follow even if the condition does not hold. In order to explore the ideas further, consider some examples.

Examples:

(1) 'A person is a husband *only if* the person is male.' This is an example of a necessary condition. The prerequisite, or necessary condition, for being a husband is that the person must be male. The

statement 'The person is a husband' cannot be true unless the statement 'The person is male' is true. If the first statement is called p and the second statement q, then p is true *only if* q is true. Yet another way of expressing the relationship is to say that 'p implies q'. Thus we have three different ways of saying the same thing:

q is a necessary condition for p
p is true only if q is true
p implies q

Also note that the reverse relationship is *not* true: to be a husband is not a necessary condition for being a male. Another way of thinking of the relationship between the two is to use the concepts introduced in the previous section, and to note that husbands form a subset of the set of all male persons.

(2) 'A person is a university graduate *if* he has a bachelor's degree awarded by the University of Bristol.' Here we have a different situation, which again can be described in three equivalent ways:

p is true *if* q is true
q is a *sufficient* condition for p
q implies p

But note that a Bristol degree is not a prerequisite for a university graduate; in other words q is sufficient for p but q is *not* a necessary condition for p. In set language, the set of university graduates contains the set of those who hold Bristol degrees; or, to put it another way, those who hold Bristol degrees are a subset of all university graduates.

(3) 'This is the first day of the (calendar) year *if and only if* the date is January 1st.' This is a third type of situation in which q is both necessary and sufficient for p. Here p implies q, *and* q implies p. In set language, one set contains the other and vice-versa, and hence the two sets are identical. (Incidentally, 'if and only if' is often contracted and written as 'iff'.)

Often the logical situation to be examined is somewhat more complicated. In the above examples, a *single* condition (q) was necessary, sufficient, or necessary and sufficient for the condition p. But often the result depends on more than one condition; the next example illustrates.

Example:

(4) Suppose x is a number drawn from the series of all non-negative integers 0, 1, 2, 3, ... Now consider the statements

$$p: x=1$$
$$q: x<2$$
$$r: x \text{ is an odd number}$$

Clearly, q is a necessary condition for p. But q (*by itself*) is not a sufficient condition, since the value $x=0$ satisfies condition q but does not make true the condition p. Similarly r is a necessary condition for p, but r (*by itself*) is not a sufficient condition. However q and r taken together comprise sufficient conditions for p. Thus we may write $x=1$ *if and only if* $x<2$ *and* x is an odd number (remembering that x is defined initially as coming from the set of non-negative integers).

These considerations are often employed in proving theorems. In some cases it may be possible to find only a necessary condition *or* perhaps simply a sufficient condition for some result or statement. But often some condition (or set of conditions) is both necessary and sufficient for the truth of the statement in question, in which case the proof of the theorem is in two parts: it is necessary to prove that the result follows *if* the condition is met (the sufficiency part); and it is also necessary to show that the result follows *only if* the condition is met (the necessity part). Sometimes this latter part of the proof is done not directly, but in an equivalent way, namely by showing that if the condition is not met then the result does not follow.

1.12 *Exercises*

1. Given the sets $A=\{1, 2, 3, 4\}$, $B=\{3, 6\}$ and $C=\{4, 5\}$ find

   (a) $A\cup B\cup C$ (b) $A\cap B$ (c) $A\cap C$
   (d) $B\cap C$ *(e) $(A\cap B)\cup C$ *(f) $A\cap(B\cup C)$

2. Which of the following statements are valid? Give reasons.

   (a) $A\cup A=A$ *(b) $A\cap A=A$
   *(c) $A\cup\varnothing=A$ (d) $A\cap\varnothing=A$

3. For the set $\{a, b, c\}$ enumerate all proper subsets having at least one element.
4. The sets previously considered have each had a finite number of

elements. But it is not necessary to limit the set concept in this way. For example, we might be interested in all values of the *continuous* variable x which are greater than 10, i.e. not only integers such as 11, 12 etc. but all the intermediate values such as 11·28, and so forth. This set has an infinite number of values. The usual notation for this kind of situation is to write, for example, $S = \{x | x > 10\}$. More generally the set is described, with the variable or variables appearing before the vertical bar, and a description of the range of values written after the bar. (This notation may also be used for a set having a finite number of elements.) As an exercise in this area, consider the sets $A = \{x | x \geqq 1\}$, $B = \{x | x \leqq 4\}$, $C = \{x | 0 \leqq x \leqq 2\}$, $D = \{x | x \leqq 1\}$ and $E = \{x | x \leqq 0\}$, and find (perhaps by considering a graph indicating the sets)

*(a) $A \cap B$ *(b) $A \cup B$ (c) $A \cap C$
(d) $A \cup C$ (e) $A \cap D$ (f) $A \cap E$

5. In each of the following cases, use the concepts of necessary and sufficient conditions to discuss the relationship between statement p on the one hand, and statement q (and statement r, when there is one) on the other hand.

*(a) p: this geometric figure has four sides
q: it is a square
(b) p: this figure is a square
q: it has four equal sides, each of which is perpendicular to the adjacent sides
(c) p: that animal has four legs
q: that animal is a horse

(For the following cases, you need to know that in Ruritania, the government pays an old-age pension to every male aged 65 or over who is not in paid employment.)

(d) p: John is eligible for a pension
q: John has already had his 65th birthday
*(e) p: John is eligible for a pension
q: John is at least 65 years of age
r: John is not in paid employment

(f) p: John is eligible for a pension
q: John is aged 67
r: John is not in paid employment

(Hint: in this last case, beware. Are q and r sufficient for p? Are they *also* necessary?)

6. Find a set of sufficient conditions on the values of x and y for the following inequality to hold:

$$(x-3)(y+4) > 0$$

Are there any alternative sets of sufficient conditions? Can you find any necessary conditions? Prepare a graph (with values of x and y on the axes) to illustrate your answer.

CHAPTER 2

Vectors

2.1 *Vectors – the basic concepts*

We have already seen that we need to handle many equations involving many variables. The concept of a vector is introduced now to help us to analyse such systems and to provide a compact notation.

In two-dimensional space, the coordinates of a point may be written (x_1, x_2). A point in three-dimensional space could be described in terms of its coordinates $(x_1, x_2, x_3) = (6, 4, 7\frac{1}{2})$ or whatever the particular numerical values might be. And we can generalize all this to the case of n dimensions. In each such case we are dealing with an *ordered* array of numbers. In other words, the order matters; the point (4, 3) is not the same as the point (3, 4). Such ordered arrays arise in many other circumstances. For example, the baker in our example in section 1.8 had stocks comprising 15 lb. of flour, 18 oz. of sugar and 20 oz. of butter. If we had many such lists to deal with, we would find it convenient to write out simply the numbers, putting them in the proper order, e.g. (15, 18, 20). By its position in first place, we know that the 15 refers to the quantity of flour measured in lb.; the 18 in second place refers to the weight of sugar in oz., and so forth; this follows from the convention we have set up to list them always in this order and in the stated units. The usefulness of such lists or ordered arrays leads us to create a formal concept, called a vector:

Definition. An n-component vector $\mathbf{a}$ is an ordered n-tuple of numbers written as a column:

$$\mathbf{a} = \begin{bmatrix} a_1 \\ a_2 \\ \cdot \\ \cdot \\ \cdot \\ \cdot \\ a_n \end{bmatrix}$$

Alternatively it may be written as a row, in which case we place a prime after the letter used to denote the vector:

$$\mathbf{a}' = [a_1\, a_2 \ldots . a_n]$$

Bold lower-case letters such as **a** and **a**′ will be used to denote vectors, to distinguish them from ordinary variables or *scalars* (as they are often called). Thus a scalar is a single variable or a number (in contrast to an ordered array) and will be denoted as hitherto by italic type; a vector is an ordered array and will be denoted by bold type. (Lower-case, i.e. small letters, are used for vectors; capital letters are used for another concept, to be introduced in Chapter 3.) A common convention (which is not universally used, however) is to denote the elements or components of a vector by subscripted variables using the same lower-case letter. Thus, as in the definition above, the elements of **a** are denoted $a_1, a_2, \ldots, a_n$; similarly we might write a vector of order 4 (i.e. having 4 components or elements) as $\mathbf{b}' = [b_1 \quad b_2 \quad b_3 \quad b_4]$. Remember that lower-case bold letters without a prime are used for column vectors, and such letters with a prime for row vectors. (As will be seen shortly, it is necessary to distinguish between a row vector having certain elements and a column vector with identical elements; this distinction has been made above by the use of the prime notation, and it is necessary to keep it in mind in carrying out the vector operations introduced in the next section.)

Examples:

(1) $[2, 4, -6]$ is a row vector of order 3

(2) $\begin{bmatrix} c_1 \\ c_2 \end{bmatrix} = \mathbf{c}$ is a column vector of order 2

Definition. The ith *unit vector* has unity as the value of its ith component, and zero value for every other component.

For vectors of order 2, there are thus two unit vectors, [1, 0] and [0, 1] which are called, respectively, the first and second unit vectors of order 2. Similarly for vectors of order n, there is a set of n different unit vectors, $[1, 0, 0, \ldots, 0]$, $[0, 1, 0, \ldots, 0]$ and so forth. Often the ith unit vector (of whatever order is implied by the context) is denoted $\mathbf{e}_i$; unfortunately this contradicts the usual notational rules.

Definition. The *null vector* **0** is a vector having all its elements equal to zero.

Believe it or not, this concept is useful, as will be seen later.

2.2 *Vector operations*

In arithmetic and in elementary algebra (or *scalar* algebra, to use a more precise term), there are well-known rules for the common operations of addition, subtraction, etc. The next task is to *define* these operations for our new concept, the vector. Note that it *is* a matter of definition: we are free to choose any consistent set of rules. Clearly it makes sense to choose rules which are useful for analysis and for computation.

Suppose that our baker got in touch with another baker and found that the latter's stocks were $\mathbf{b}' = [35, 10, 20]$, where this vector representation is based on the same convention as previously used. In other words, the second baker has stocks of 35 lb of flour, 10 oz. of sugar and 20 oz. of butter. How should we associate this new vector with the original vector of $\mathbf{a}' = [15, 18, 20]$?

The obvious answer is to associate *corresponding* elements or components. Thus, for example, the third component in each case relates to the stock of butter measured in oz.; in fact, in the case of this commodity, the bakers have stocks of equal size. It does not make sense to associate the first component of **a** with the second element of **b**; one relates to flour and the other to sugar, and the quantities cannot be compared, or added together, or otherwise handled together. This idea of associating each element of a vector with the corresponding element of the other vector leads us to useful definitions of our first few concepts and operations:

Definition. Two vectors **a**, **b** each of order n are said to be *equal*, i.e. $\mathbf{a} = \mathbf{b}$ if all the corresponding components are equal, i.e. if $a_i = b_i \quad i = 1, 2, \ldots, n$

This implies that two vectors cannot be compared for equality unless they have the same number of elements.

The concept of vector inequality gives a little more trouble.

Definition. Given two vectors **a**, **b** each of order n, then

$$\mathbf{a} > \mathbf{b} \quad \text{if } a_i > b_i \quad i = 1, 2, \ldots, n$$

and

$$\mathbf{a} \geqq \mathbf{b} \quad \text{if } a_i \geqq b_i \quad i = 1, 2, \ldots, n$$

In the first case, $a_i > b_i$ for *all* components. In the second case, commonly some $a_i > b_i$ while some components are equal. The second case can include the circumstances where all corresponding components are equal. Sometimes we may wish to exclude this; in other words we may wish to describe the requirement that all $a_i \geqq b_i$ with at least one $a_i > b_i$ (without specifying in advance which element or elements must be unequal). One notation for this is to write $\mathbf{a} \geq \mathbf{b}$ but this convention is not universally employed.

The concept of addition is quite straightforward. We can only add like to like, i.e. add corresponding elements:

Definition. The sum of two n-component vectors $\mathbf{a}$ and $\mathbf{b}$ is written $\mathbf{a}+\mathbf{b}$ and is defined

$$\mathbf{a}+\mathbf{b} = \begin{bmatrix} a_1+b_1 \\ a_2+b_2 \\ \cdot \\ \cdot \\ \cdot \\ \cdot \\ a_n+b_n \end{bmatrix}$$

(Alternatively and equivalently, this definition could be written out for row vectors.) Again, this applies only to vectors having the same number of elements, and the result is a vector of the same order.

Definition. The product of a scalar λ and an n-component vector $\mathbf{a}$ is defined as

$$\lambda\mathbf{a}' = [\lambda a_1 \; \lambda a_2 \; \lambda a_3 \; \ldots . \; \lambda a_n]$$

(where this time the definition has been written out in the form for row vectors).

This permits a definition of subtraction in terms of the operations of addition and of multiplication by a scalar:

Definition. For two n-component vectors $\mathbf{a}$ and $\mathbf{b}$

$$\mathbf{a}-\mathbf{b} = \mathbf{a}+(-1)\mathbf{b} = \begin{bmatrix} a_1-b_1 \\ a_2-b_2 \\ \cdot \\ \cdot \\ \cdot \\ \cdot \\ a_n-b_n \end{bmatrix}$$

Examples:

(1) $\begin{bmatrix} 2 \\ 3 \end{bmatrix} + \begin{bmatrix} 4 \\ 7 \end{bmatrix} = \begin{bmatrix} 6 \\ 10 \end{bmatrix}$

(2) $[2, 3] + [4, 7] = [6, 10]$

(3) $[3, 6] > [2, 5]$

(4) $[2, 6] \leqq [3, 6]$

(5) $2\begin{bmatrix} 3 \\ 17 \end{bmatrix} = \begin{bmatrix} 6 \\ 34 \end{bmatrix}$

The operations defined in this section have the usual associative, commutative and distributive properties which are found in the elementary algebra of scalars. That is to say, the following properties hold:

(1) $\mathbf{a}+(\mathbf{b}+\mathbf{c}) = (\mathbf{a}+\mathbf{b})+\mathbf{c}$

(2) $\mathbf{a}+\mathbf{b} = \mathbf{b}+\mathbf{a}$

(3) $(\lambda+\phi)\mathbf{a} = \lambda\mathbf{a}+\phi\mathbf{a}$

(4) $\lambda(\mathbf{a}+\mathbf{b}) = \lambda\mathbf{a}+\lambda\mathbf{b}$

where **a**, **b** and **c** are vectors of order n and where λ and ϕ are scalars. Each of these properties can be proved in a straightforward way. Consider property (1) – the brackets are used to denote the order in which the addition operations are carried out.

First $(\mathbf{b}+\mathbf{c}) = \begin{bmatrix} b_1+c_1 \\ b_2+c_2 \\ \cdot \\ \cdot \\ \cdot \\ \cdot \\ b_n+c_n \end{bmatrix}$ and then $\mathbf{a}+(\mathbf{b}+\mathbf{c}) = \begin{bmatrix} a_1+b_1+c_1 \\ a_2+b_2+c_2 \\ \cdot \\ \cdot \\ \cdot \\ \cdot \\ a_n+b_n+c_n \end{bmatrix}$

Also $(\mathbf{a}+\mathbf{b}) = \begin{bmatrix} a_1+b_1 \\ a_2+b_2 \\ \cdot \\ \cdot \\ \cdot \\ \cdot \\ a_n+b_n \end{bmatrix}$ and then $(\mathbf{a}+\mathbf{b})+\mathbf{c} = \begin{bmatrix} a_1+b_1+c_1 \\ a_2+b_2+c_2 \\ \cdot \\ \cdot \\ \cdot \\ \cdot \\ a_n+b_n+c_n \end{bmatrix}$

Thus either route gives the same result.

As an exercise, use similar methods to derive the other three properties.

We have not considered here the multiplication of one vector into another vector; this will be treated in a later chapter; see also exercise 5 in the next section.

2.3 *Exercises*

1. If $\mathbf{a}' = [4, 0, 2, 7]$, $\mathbf{b}' = [-8, 2, 1, 6]$ and $\mathbf{c}' = [4, -2, 6]$ consider whether each of the following operations is defined, and where it is, carry out the appropriate calculation:

   (a) $\mathbf{a}+\mathbf{b}$ (b) $\mathbf{a}-\mathbf{c}$ (c) $\mathbf{a}'-2\mathbf{b}'$
   *(d) is $\mathbf{a} > \mathbf{b}$? (e) is $\mathbf{a} > \mathbf{c}$?

*2. In an ordinary economic system, all prices p_i $(i = 1, 2, \ldots, n)$ are non-negative, i.e. they are either positive or zero. If $\boldsymbol{p}$ denotes the vector of such prices, express this non-negativity property in vector notation, carefully describing each magnitude as either a scalar or a vector, and stating the order in the case of each vector.

3. A very bureaucratic organization has three grades of managers and each grade gets a fixed set of office furnishings. A senior grade manager gets a large desk, two chairs, an easy chair and a carpet; a middle grade manager gets a small desk, two chairs and one easy chair; and a junior grade manager gets a small desk and two chairs. A new branch office is to be set up with one senior grade, five middle grade and eleven junior grade managers. By setting out your calculations using vector concepts wherever possible, find the total amounts of furnishings required for the new office.

*4. Referring to exercise 3, and thinking again in vector terms, in what sense (if any) can it be said that a senior grade manager is given a 'greater' amount of furnishings than a middle grade manager? How does this relationship compare with that between a middle and junior grade manager?

*5. Suppose that flour is worth 6 pence per lb., sugar is worth 1 penny per oz., and butter is worth 2 pence per oz. Consider these prices as a vector; also consider as a vector the baker's stocks of 15 lb. of flour, 18 oz. of sugar and 20 oz. of butter. Carry out the usual calculation to find the total value of the baker's stocks. What does this suggest for a useful definition for the operation of multiplying one vector into another vector?

2.4 *Vector representation of simultaneous equations*

As already indicated, one purpose in introducing the concept of a vector is to assist in the handling of large systems of simultaneous (linear) equations. We need some further concepts before we can discuss general methods of solution, and these will be introduced later. But at this stage it is possible to show how vector notation enables such sets of equations to be written in a more compact way.

Consider a general set of m linear equations in n variables (the x_i):

$$\begin{array}{l} a_{11}x_1+a_{12}x_2+\ldots+a_{1n}x_n = b_1 \\ a_{21}x_1+a_{22}x_2+\ldots+a_{2n}x_n = b_2 \\ \cdot \\ \cdot \\ \cdot \\ \cdot \\ a_{m1}x_1+a_{m2}x_2+\ldots+a_{mn}x_n = b_m \end{array}$$

Note that the coefficients a_{11}, a_{21} etc. appearing in the first term in each equation are multiplied by the same variable x_1, which is of course a scalar. This suggests that we might employ the operation of multiplying a vector by a scalar. Thus the equations may be written

$$x_1\begin{bmatrix} a_{11} \\ a_{21} \\ \cdot \\ \cdot \\ \cdot \\ \cdot \\ a_{m1} \end{bmatrix} + x_2\begin{bmatrix} a_{12} \\ a_{22} \\ \cdot \\ \cdot \\ \cdot \\ \cdot \\ a_{m2} \end{bmatrix} + \ldots + x_n\begin{bmatrix} a_{1n} \\ a_{2n} \\ \cdot \\ \cdot \\ \cdot \\ \cdot \\ a_{mn} \end{bmatrix} = \begin{bmatrix} b_1 \\ b_2 \\ \cdot \\ \cdot \\ \cdot \\ \cdot \\ b_m \end{bmatrix}$$

Check for yourself that this *vector equation* is equivalent to the original set of equations.

Incidentally, this representation suggests one use for the null vector. If the equations all have zero on the right-hand side, i.e. if all b_i for $i = 1, 2, \ldots, m$ are zero, then we need a null vector on the right-hand side of the vector representation.

2.5 *Euclidean space and linear dependence*

It is now time to develop some further vector concepts which are based on the vector operations already introduced. The notational convention introduced in section 2.1 will be broken here, since it is sometimes convenient to use subscripted variables to represent *vectors*. In particular, we will use $\mathbf{a}_1$, $\mathbf{a}_2$ etc. to denote members of a set of vectors; these are then put in bold type, of course. At the same time, we will still use subscripted variables in italics (e.g. b_1, b_2 etc.) to represent scalars which are (or can be regarded as) elements of a vector.

Definition. Given a set (m in number) of n-component vectors denoted $\mathbf{a}_1, \mathbf{a}_2, \ldots. \mathbf{a}_m$, the n-component vector

$$\mathbf{a} = \sum_{i=1}^{m} \lambda_i \mathbf{a}_i = \lambda_1 \mathbf{a}_1 + \lambda_2 \mathbf{a}_2 + \ldots + \lambda_m \mathbf{a}_m$$

is called a *linear combination* of the vectors $\mathbf{a}_1, \ldots, \mathbf{a}_m$, where the scalars λ_i $(i = 1, \ldots, m)$ may take on any value.

Examples:

(1) If $\mathbf{a}_1 = \begin{bmatrix} 5 \\ 7 \end{bmatrix}$ and $\mathbf{a}_2 = \begin{bmatrix} 2 \\ -1 \end{bmatrix}$

any linear combination of $\mathbf{a}_1$ and $\mathbf{a}_2$ may be called $\mathbf{a}$, and written

$$\mathbf{a} = \lambda_1 \mathbf{a}_1 + \lambda_2 \mathbf{a}_2 = \lambda_1 \begin{bmatrix} 5 \\ 7 \end{bmatrix} + \lambda_2 \begin{bmatrix} 2 \\ -1 \end{bmatrix}$$

$$= \begin{bmatrix} 5\lambda_1 + 2\lambda_2 \\ 7\lambda_1 - \lambda_2 \end{bmatrix}$$

The various possible values for the components of $\mathbf{a}$ are generated as λ_1 and λ_2 take on all possible values.

If $\lambda_1 = 2$ and $\lambda_2 = -3$, for example, then

$$\mathbf{a} = \begin{bmatrix} 4 \\ 17 \end{bmatrix}$$

(2) One particular kind of application of the linear combination idea is of especial importance. This arises when we consider linear combinations of unit vectors. To be specific, consider the entire set of n such vectors, each of which has n components of course, and let these vectors be denoted (as before) by $\mathbf{e}_1$, $\mathbf{e}_2$, $\ldots$, $\mathbf{e}_n$. Now *any* n-component vector can be expressed as a

linear combination of this set, as the following example shows. Take any vector of order n and let it be denoted

$$\mathbf{b} = \begin{bmatrix} b_1 \\ b_2 \\ \cdot \\ \cdot \\ \cdot \\ \cdot \\ b_n \end{bmatrix}$$

(Note that b_1, b_2 etc. are here used to represent scalars.) Then we may write

$$\mathbf{b} = b_1\mathbf{e}_1 + b_2\mathbf{e}_2 + \ldots . + b_n\mathbf{e}_n$$

Check for yourself that this vector equation is valid.

(3) The vector $\mathbf{a}' = [3, 13, -4]$ may be expressed as a linear combination of unit vectors:

$$\begin{aligned} \mathbf{a} &= 3\mathbf{e}_1 + 13\mathbf{e}_2 - 4\mathbf{e}_3 \\ &= 3\begin{bmatrix} 1 \\ 0 \\ 0 \end{bmatrix} + 13\begin{bmatrix} 0 \\ 1 \\ 0 \end{bmatrix} - 4\begin{bmatrix} 0 \\ 0 \\ 1 \end{bmatrix} \end{aligned}$$

Thus we may use the unit vectors to define the coordinate system for our n-dimensional space; each component of a vector measures one of the coordinates of a point in n-dimensional space. This is a natural extension of the graphical procedures in Chapter 1 where the coordinates of a point corresponded to the components of what were (in effect) two-dimensional vectors. A formal definition of this n-dimensional space is convenient:

Definition. An *n-dimensional Euclidean space* is defined as the collection of all n-component vectors.

Such a space is usually denoted E^n. Our elementary intuitive notions of two-dimensional space and of three-dimensional space are simply examples of this more general concept. As an alternative to speaking of n-component vectors, we can refer to vectors from E^n.

If a set of vectors is selected from E^n, it may be possible, for this particular set, to express one vector as a linear combination of the other vectors. If this is possible, we can in some sense drop that vector from the set without great loss; the precise sense will become apparent in our later discussion of simultaneous equations. If it is not possible to express one vector as a linear combination of the others, all vectors in the set are in some sense independent of the rest of the set. We will formalize this by introducing a definition of linear dependence, which is then shown to be related to these matters.

Definition. A set of vectors $\mathbf{a}_1, \mathbf{a}_2, \ldots, \mathbf{a}_m$ from E^n is said to be *linearly dependent* if there exists a set of scalars λ_i not all zero and such that

$$\lambda_1\mathbf{a}_1 + \lambda_2\mathbf{a}_2 + \ldots + \lambda_m\mathbf{a}_m = \mathbf{0}$$

(Note that the right-hand side comprises the null vector, of order n.)

On the other hand, if the only set of λ_i for which this equation holds is

$$\lambda_1 = \lambda_2 = \ldots = \lambda_m = 0$$

then the vectors $\mathbf{a}_i$ are said to be *linearly independent*.

These definitions imply that a set of vectors which is not linearly dependent must be linearly independent, and vice-versa. In other words the two categories are mutually exclusive, and exhaustive (i.e. there are no other categories with respect to this property).

Notice that the definition is not in the same terms as expressing 'one vector as a linear combination of the other vectors'. However, the two forms of expression are equivalent, as will now be proved.

Theorem 2.1. The vectors $\mathbf{a}_1, \mathbf{a}_2, \ldots, \mathbf{a}_m$ from E^n are linearly dependent if and only if it is possible to find at least one vector in the set such that it can be expressed as a linear combination of the remaining vectors.

Proof (a). First the sufficiency condition will be proved; in other words, it will be shown that linear dependence follows *if* one vector can be expressed as a linear combination of the others. Suppose (without loss of generality) that the vector in question is numbered as $\mathbf{a}_1$. Then consider its expression as a linear combination of the other vectors:

$$\mathbf{a}_1 = \lambda_2\mathbf{a}_2 + \lambda_3\mathbf{a}_3 + \ldots + \lambda_m\mathbf{a}_m$$

i.e.

$$(-1)\,\mathbf{a}_1 + \lambda_2\mathbf{a}_2 + \ldots + \lambda_m\mathbf{a}_m = \mathbf{0}$$

But this satisfies the condition for linear dependence since the first coefficient (at the least) is not zero.

Proof (b). Next the necessity condition will be established; in other words it will be shown that linear dependence follows *only if* one vector can be expressed as a linear combination of the others. This will be proved by showing that if the vectors are linearly dependent it is always possible to express one as a linear combination of the others; equivalently this means that linear dependence necessarily implies linear combination, i.e. cannot occur unless expression as a linear combination is possible. (This indirect form of proof is often used to establish necessary conditions.)

Since (by supposition) the vectors are linearly dependent,

$$\lambda_1\mathbf{a}_1 + \lambda_2\mathbf{a}_2 + \ldots . + \lambda_m\mathbf{a}_m = \mathbf{0}$$

with at least one $\lambda_i \neq 0$. Suppose the λ_i are numbered such that $\lambda_1 \neq 0$. Then both sides of the equation may be divided by λ_1, and terms rearranged to make $\mathbf{a}_1$ the subject of the equation:

$$\mathbf{a}_1 = -\frac{\lambda_2}{\lambda_1}\mathbf{a}_2 - \ldots . - \frac{\lambda_m}{\lambda_1}\mathbf{a}_m$$

Thus one vector has been written as a linear combination of the other vectors.

This proof not only establishes the equivalence of linear dependence of a set of vectors and the ability to express one vector as a linear combination of the others, but also gives some practice in the methods of formulation and proof of necessary and sufficient conditions.

Examples:

(1) Consider the vectors $\begin{bmatrix} 2 \\ 5 \end{bmatrix}$ and $\begin{bmatrix} 1 \\ 3 \end{bmatrix}$

Now suppose there exist scalars λ_i such that

$$\lambda_1\begin{bmatrix} 2 \\ 5 \end{bmatrix} + \lambda_2\begin{bmatrix} 1 \\ 3 \end{bmatrix} = \mathbf{0}$$

This yields the pair of equations

$$\begin{aligned} 2\lambda_1 + \lambda_2 &= 0 \\ 5\lambda_1 + 3\lambda_2 &= 0 \end{aligned}$$

which has as the only solution

$$\lambda_1 = 0 \quad \text{and} \quad \lambda_2 = 0$$

Thus the two vectors are a linearly independent set. (Since a solution exists for the λ_i, the supposition is justified.)

(2) The vectors $\begin{bmatrix} 2 \\ 4 \end{bmatrix}$ and $\begin{bmatrix} 4 \\ 8 \end{bmatrix}$ are linearly dependent. One way of seeing this is to observe that

$$\begin{bmatrix} 4 \\ 8 \end{bmatrix} = 2 \times \begin{bmatrix} 2 \\ 4 \end{bmatrix}$$

Three further theorems are stated here without proof. (Each may be proved in an elementary fashion, by simply applying the definitions given above.)

Theorem 2.2. If a set of vectors is linearly dependent, any larger set containing the initial set is also linearly dependent.

Theorem 2.3. If a set of vectors is linearly independent, any subset of these vectors is also linearly independent.

Theorem 2.4. Given a set of vectors $\mathbf{a}_1, \mathbf{a}_2, \ldots, \mathbf{a}_m$ from E^n, consider all subsets of this set, and suppose that the largest subset which is linearly independent has k vectors (where $k<m$). Then given any such linearly independent subset, each other vector in the main set can be expressed as a linear combination of the k vectors which form the subset.

For this last theorem, note that the result follows not for *any* linearly independent subset but only for those which comprise k vectors, where k is the number of vectors in the largest linearly independent subset.

In very small examples, it is usually not too difficult to find out whether or not a set of vectors is linearly dependent. In other cases, however, the computation can be quite tricky; fortunately there are some formal methods for doing this, and these will be discussed later in Chapter 4.

2.6 *Exercises*

*1. Express [3, 14] as a linear combination of [3, 4] and [1, −2].

2. Show that [8, 0], [5, 6] and [2, −4] are linearly dependent.

3. Express $\begin{bmatrix} 5 \\ 6 \\ 8 \end{bmatrix}$ as a linear combination of $\begin{bmatrix} 1 \\ 2 \\ 3 \end{bmatrix}$, $\begin{bmatrix} 4 \\ 5 \\ 6 \end{bmatrix}$ and $\begin{bmatrix} 0 \\ 1 \\ 1 \end{bmatrix}$.

4. Show that $\begin{bmatrix} 1 \\ 0 \\ 1 \end{bmatrix}$, $\begin{bmatrix} 2 \\ 1 \\ 0 \end{bmatrix}$ and $\begin{bmatrix} 0 \\ 1 \\ 1 \end{bmatrix}$ are linearly independent.

*5. Show that a set of n-component vectors which includes the null vector must be linearly dependent.

*6. Consider a set of vectors comprising a single vector. Under what circumstances (if any) is this set linearly independent?

7. If all prices are non-negative, do all possible price vectors comprise a Euclidean space?

2.7 *Bases*

It was observed in section 2.5 that any n-component vector may be written as a linear combination of the set of n unit vectors. More generally, we are interested not just in sets of unit vectors but in *all* sets of vectors which can be used in this way.

Definition. A set of vectors $\mathbf{a}_1, \mathbf{a}_2, \ldots, \mathbf{a}_m$ from E^n is said to *span* E^n if every vector in E^n can be written as a linear combination of this set of vectors (which is then called a *spanning set*).

Now consider devising such a set with as few vectors in it as possible. If the spanning set is linearly dependent, then at least one vector in it can be expressed as a linear combination of the others. To show that this property can be used to reduce the size of the spanning set, consider an example in E^2. Note that

$$\mathbf{a} = \begin{bmatrix} a_1+2 \\ a_2+2 \end{bmatrix}$$

represents any vector in E^2; in other words the scalars a_1 and a_2 can take on any values, and hence any vector can be represented in this way. Now $\mathbf{a}$ may be written as:

$$\mathbf{a} = a_1\begin{bmatrix} 1 \\ 0 \end{bmatrix} + a_2\begin{bmatrix} 0 \\ 1 \end{bmatrix} + 2\begin{bmatrix} 1 \\ 1 \end{bmatrix}$$

and thus these three vectors span E^2. But they are a linearly dependent set since

$$\begin{bmatrix}1\\1\end{bmatrix}=\begin{bmatrix}1\\0\end{bmatrix}+\begin{bmatrix}0\\1\end{bmatrix}$$

Obviously this can be substituted into the expression for **a**:

$$\begin{aligned}\mathbf{a} &= a_1\begin{bmatrix}1\\0\end{bmatrix}+a_2\begin{bmatrix}0\\1\end{bmatrix}+2\begin{bmatrix}1\\0\end{bmatrix}+2\begin{bmatrix}0\\1\end{bmatrix}\\ &= (a_1+2)\begin{bmatrix}1\\0\end{bmatrix}+(a_2+2)\begin{bmatrix}0\\1\end{bmatrix}\end{aligned}$$

In this case, we have shown that a linearly dependent spanning set can be reduced in size; in fact the spanning set which remains in this particular instance is linearly independent (since it comprises unit vectors which may readily be seen to have this property – check this for yourself). More generally the same kind of algebraic manipulation can always be applied; thus a linearly dependent spanning set can always be reduced (by a series of substitutions, if necessary) to an independent set. This provides the motivation for our next definition.

Definition. A *basis* for E^n is a linearly independent set of vectors from E^n which spans the entire space.

Note that a basis for E^n is not unique. In E^2 for example, the unit vectors form a basis (as they do in any Euclidean space); a sample of other bases are [2, 0] and [0, 1]; [4, 0] and [0, 4]; and [1, 1] and [2, 3]. The first two pairs of vectors are obviously bases since they are close relatives of the unit vectors.

To show that the third pair is indeed a basis, let us begin by supposing that scalars λ and ϕ exist such that any general vector $\mathbf{x}' = (x_1, x_2)$ can be expressed

$$\mathbf{x} = \lambda\begin{bmatrix}1\\1\end{bmatrix}+\phi\begin{bmatrix}2\\3\end{bmatrix}$$

In other words we are supposing the result to be true, and then exploring whether this leads to any difficulties in computing λ and ϕ. This expression may be written as two equations in the scalars involved:

$$x_1 = \lambda + 2\phi$$
$$x_2 = \lambda + 3\phi$$

By subtraction, $\phi = -x_1 + x_2$

and hence $\lambda = 3x_1 - 2x_2$

Thus our assumption has not led to any trouble – we can always compute scalars λ and ϕ no matter what the values of x_1 and x_2.

A similar approach (though needing greater care) can be used to show that *any* pair of *linearly independent* vectors in E^2 is a basis for E^2. The general result applies in all Euclidean spaces, and is now stated formally without proof.

Theorem 2.5. Any set of n linearly independent vectors from E^n forms a basis for E^n.

This result can be used to prove the next property.

Theorem 2.6. Any set of $n+1$ vectors from E^n is linearly dependent.

Proof. Begin by supposing that the theorem does not hold. If then the $n+1$ vectors are linearly independent, the first n vectors in the list must also be linearly independent (from Theorem 2.3).

Thus these n vectors form a basis (from Theorem 2.5). Hence the remaining vector can be expressed as a linear combination of this basis set, and thus the $n+1$ vectors are linearly dependent. Thus the initial supposition has led to a contradiction, and hence it must be wrong; in other words, the theorem is valid.

Theorem 2.7. For a given set of basis vectors, any vector in E^n can be written as a linear combination of these vectors in only one way.

Proof. Suppose that the theorem is not true i.e. suppose any vector $\mathbf{b}$ can be written in two distinct ways in terms of the basis set:

$$\mathbf{b} = \lambda_1\mathbf{a}_1 + \lambda_2\mathbf{a}_2 + \ldots\ldots + \lambda_m\mathbf{a}_m$$
$$\mathbf{b} = \phi_1\mathbf{a}_1 + \phi_2\mathbf{a}_2 + \ldots\ldots + \phi_m\mathbf{a}_m$$

where the scalars λ_i and ϕ_i differ from each other for at least one value of i.

If one equation is subtracted from the other (remembering our definition for the subtraction of vectors), we obtain

$$(\lambda_1 - \phi_1)\,\mathbf{a}_1 + \ldots\ldots + (\lambda_m - \phi_m)\,\mathbf{a}_m = \mathbf{0}$$

Since at least one of the coefficients $\lambda_i - \phi_i$ is non-zero, this leads to the conclusion that the basis vectors are linearly dependent. But, by definition, a basis comprises linearly independent vectors. Thus our initial supposition (that **b** can be written in two distinct ways) has led to a contradiction and must therefore be wrong. In other words, there is a unique expression for **b** in terms of the basis vectors, and the theorem is proved.

2.8 *Exercises*

1. Does the set of vectors [3, 0, 1], [7, 0, 9] and [3, 1, 1] form a basis for E^3?

*2. Is the set of vectors $\begin{bmatrix} 2 \\ 1 \end{bmatrix}$, $\begin{bmatrix} 1 \\ 2 \end{bmatrix}$ and $\begin{bmatrix} 0 \\ 3 \end{bmatrix}$ a spanning set for E^2? Is it also a basis?

3. If θ may take on any positive value whatsoever, what can be said about the linear dependence or independence of the set of vectors $[1, \theta]$, $[\theta, 1]$ and $[\theta, 2\theta]$?

*4. If vectors **a**, **b**, and **c** from E^n are linearly independent, show that the set of vectors $\mathbf{a}+\mathbf{b}$, $\mathbf{b}+\mathbf{c}$ and $\mathbf{c}+\mathbf{a}$ is also linearly independent.

*5. For what value of θ (if any) is the set of vectors [1, 2] and $[4, \theta]$ not a basis?

*6. A firm is considering undertaking any one of three investment schemes, denoted by *A*, *B* and *C*. Each scheme involves an initial expenditure of £380,000 in year 0; each investment will earn the receipts shown (in the following table) in years 1, 2 and 3; after that there are no further receipts.

Receipts in each year (£000)

	A	*B*	*C*
Year 1	100	150	100
Year 2	150	140	150
Year 3	200	130	210

The firm notices that *B* earns receipts which decrease as time passes, while for the others the opposite is the case. The firm realizes that £1 received in one year is not worth the same as £1 in another year, and decides to carry out some interest calculations in order to make a detailed comparison. But before this is done, what (if anything) can be said about the comparison between:

(i) scheme A and scheme B
(ii) scheme A and scheme C?

CHAPTER 3

Matrices

3.1 *The concept of a matrix*

A scalar is a single number. A vector is a list or array of scalars in one dimension, so to speak; in other words, a series of scalars is arranged one after the other; in a column vector there is only one scalar in each row (and vice-versa). We now introduce a generalization of a vector in which the array of scalars is in *two* dimensions:

Definition. A matrix is a rectangular array of numbers arranged in rows and columns. More particularly, a matrix of order $m \times n$ is a set of $m \times n$ elements arranged in m rows and n columns.

Notation for a matrix is not always standardized. The notation used here is one which commonly employed. A matrix of order $m \times n$ can be written in full:

$$\begin{bmatrix} a_{11} & a_{12} & \cdots & a_{1n} \\ a_{21} & a_{22} & \cdots & a_{2n} \\ \cdot & & & \\ \cdot & & & \\ \cdot & & & \\ \cdot & & & \\ a_{m1} & a_{m2} & \cdots & a_{mn} \end{bmatrix}$$

Thus square brackets are used to enclose the array of elements; where the matrix is of general variables rather than of particular numerical values, each of the elements is represented by a double-subscripted variable; the element in the ith row and j^{th} column has subscripts i and j (in that order). For a short-hand notation, an upper-case letter (here, **A**) can stand for the matrix; sometimes the matrix is represented by a typical element, enclosed in square brackets:

$$\mathbf{A} = [a_{ij}]$$

(Notice that this use of double-subscripted variables is the same as that introduced in section 1.3 for the elements of a table. And as with vectors, the letter denoting a matrix is printed in bold type.)

A matrix can have any relation between m (the number of rows) and n (the number of columns). Thus we can have $m > n$ or $m < n$, for example, giving what are called *rectangular* matrices. If $m = 1$, the matrix reduces to a row vector; if $n = 1$, it is equivalent to a column vector. It is in this sense that a matrix is a generalization of a vector. (And, as will be seen below, it is sometimes useful to think of a matrix as a collection of column vectors, *or* as a collection of row vectors.) If $m = n$, the matrix is termed a *square* matrix; such a matrix with n rows (and hence n columns) is often called an nth-order matrix. If $\mathbf{A}$ is a square matrix, the set of elements $\{a_{11}, a_{22}, a_{33}, \ldots, a_{nn}\}$ is called the *principal, leading* or *main diagonal.*

Examples:

$$\begin{bmatrix} 1 & 2 & 3 \\ 1 & 7 & 9 \end{bmatrix}, \begin{bmatrix} 2 & 7 \\ -3 & 0 \end{bmatrix}, \begin{bmatrix} 0 & 0 \\ 0 & 1 \end{bmatrix}, \begin{bmatrix} 3 \\ -3 \end{bmatrix}$$

are all matrices. But the following are *not* matrices since they do not have their elements arranged in rectangular arrays comprising rows and columns:

$$\begin{bmatrix} 0 & 1 \\ & 0 \end{bmatrix}, \begin{bmatrix} 2 & 3 \\ 4 & \end{bmatrix}$$

3.2 *The first matrix operations*

A matrix is merely a table of numbers. Apart from being a convenient way of recording certain types of numerical data, it has no particular value in itself unless we can define operations which may be performed on it. As with the vector operations, it is not enough to invent operations which are logically consistent; we want operations which are useful. In the present section, we deal principally with addition and subtraction where the obvious choice is just about the only conceivable definition. In a later section, the discussion of the problem of defining matrix multiplication will show that more than one definition is not only possible but potentially useful.

Definition. Two matrices $\mathbf{A}$ and $\mathbf{B}$ are defined to be equal (written

$\mathbf{A} = \mathbf{B}$) if and only if the corresponding elements are equal, i.e. if and only if $a_{ij} = b_{ij}$ for all i and j ($i = 1, \ldots, m; j = 1, \ldots, n$).

Note that two matrices must be of the same order (i.e. one must have the same number of rows as the other, and similarly for columns) before they can be said to be equal. Bearing in mind that a vector is a particular kind of matrix, we now see why it is desirable to distinguish between a row vector and a column vector even when the two have identical elements.

As far as inequality is concerned, there is the same diversity of cases as we noted in section 2.2 in dealing with vector inequality. The same notation may be employed again here:

Definition. Given two matrices $\mathbf{A}$ and $\mathbf{B}$ each of order $m \times n$, then

$$\mathbf{A} > \mathbf{B} \text{ iff } a_{ij} > b_{ij} \text{ for all } i, j$$

and

$$\mathbf{A} \geqq \mathbf{B} \text{ iff } a_{ij} \geqq b_{ij} \text{ for all } i, j$$

For addition, the obvious association of corresponding elements yields:

Definition. The sum $\mathbf{C}$ of matrices $\mathbf{A}$ and $\mathbf{B}$ (where all three matrices are of order $m \times n$) is written

$$\mathbf{C} = \mathbf{A} + \mathbf{B}$$

where elements are given by

$$c_{ij} = a_{ij} + b_{ij} \quad (\text{all } i, j)$$

Again, matrix addition is defined only where the matrices are of the same order.

Examples:

(1) $\mathbf{A} = \begin{bmatrix} 1 & 2 \\ 4 & 3 \end{bmatrix} \quad \mathbf{B} = \begin{bmatrix} 0 & 1 \\ 1 & 1 \end{bmatrix} \quad \mathbf{C} = \mathbf{A} + \mathbf{B} = \begin{bmatrix} 1 & 3 \\ 5 & 4 \end{bmatrix}$

(2) $\mathbf{A} = \begin{bmatrix} 2 & 3 \\ 7 & 9 \end{bmatrix} \quad \mathbf{B} = \begin{bmatrix} 7 & 9 & 1 \\ 8 & -2 & 5 \end{bmatrix}$

Addition is not defined.

Definition. Given a scalar λ and a matrix $\mathbf{A}$ of order $m \times n$, the operation of multiplying the matrix by the scalar is

$$\lambda\mathbf{A} = \begin{bmatrix} \lambda a_{11} & \cdots\cdots & \lambda a_{1n} \\ \vdots & & \\ \lambda a_{m1} & \cdots\cdots & \lambda a_{mn} \end{bmatrix}$$

As with vectors, subtraction can be defined as multiplication by the scalar (-1):

Definition. Given matrices **A**, **B** each of order $m \times n$,

$$\mathbf{A}-\mathbf{B} = \mathbf{A}+(-1)\mathbf{B} = [a_{ij}-b_{ij}]$$

Examples:

(3) If **A** is as defined in example (1) above, and if $\lambda = 3$,

$$\lambda\mathbf{A} = 3\begin{bmatrix} 1 & 2 \\ 4 & 3 \end{bmatrix} = \begin{bmatrix} 3 & 6 \\ 12 & 9 \end{bmatrix}$$

(4) With **A** and **B** as in example (1) above,

$$\mathbf{A}-\mathbf{B} = \begin{bmatrix} 1 & 1 \\ 3 & 2 \end{bmatrix}$$

(5) $\begin{bmatrix} 7 & 49 \\ 700 & 77 \end{bmatrix}$ may be written (more conveniently) as $7\begin{bmatrix} 1 & 7 \\ 100 & 11 \end{bmatrix}$

As with vectors (see section 2.2), certain associative, commutative and distributive properties hold. They are stated here without proof:

(i) $\mathbf{A}+(\mathbf{B}+\mathbf{C}) = (\mathbf{A}+\mathbf{B})+\mathbf{C}$
(ii) $\mathbf{A}+\mathbf{B} = \mathbf{B}+\mathbf{A}$
(iii) $\lambda(\mathbf{A}+\mathbf{B}) = \lambda\mathbf{A}+\lambda\mathbf{B}$

3.3 *Exercises*

1. In each of the following cases, consider whether addition and subtraction are defined; where they are, find **A**+**B** and **A**−**B**

(a) $\mathbf{A} = \begin{bmatrix} 2 & 7 \\ -3 & 4 \end{bmatrix}$ $\quad \mathbf{B} = \begin{bmatrix} -3 & 4 \\ 2 & 1 \end{bmatrix}$

*(b) $\mathbf{A} = [4 \quad 7]$ $\quad \mathbf{B} = \begin{bmatrix} 6 \\ -11 \end{bmatrix}$

(c) $\mathbf{A} = \begin{bmatrix} 5 & 2 \\ 7 & 4 \end{bmatrix}$ $\quad \mathbf{B} = \begin{bmatrix} 3 & \alpha \\ 5 & \beta \end{bmatrix}$

2. If the operations are defined, find $2\mathbf{A}+3\lambda\mathbf{B}$ where

$$\mathbf{A}=\begin{bmatrix} 7 & 2 \\ 4 & 8 \\ -1 & 6 \end{bmatrix} \quad \text{and} \quad \mathbf{B}=\begin{bmatrix} 1 & 0 \\ 2 & 1 \\ 0 & 0 \end{bmatrix}$$

3. In part (a) of exercise 1 above, is $\mathbf{A}>\mathbf{B}$?
4. In part (c) of exercise 1 above, what conditions (if any) can be placed on α and β to make

   (i) $\mathbf{A}>\mathbf{B}$
   (ii) $\mathbf{A}\geqq\mathbf{B}$
   *(iii) $\mathbf{A}<\mathbf{B}$?

*5. Suppose m products are sold in each of n shops. Supposing constant prices during a year, let prices in first year be denoted by $\mathbf{P}=[p_{ij}]$ where $i=1, \ldots, m$, and $j=1, \ldots, n$. Let prices in second year be similarly denoted by a matrix $\mathbf{R}$. Let quantities sold during first and second years be similarly denoted by matrices $\mathbf{Q}$ and $\mathbf{S}$. What meaning (if any) can be attached to (i) $\mathbf{P}+\mathbf{R}$, (ii) $\mathbf{P}+\mathbf{Q}$, (iii) $\mathbf{Q}+\mathbf{S}$?

3.4 *Some particular matrix types*

Certain special matrices occur fairly frequently in both pure analysis, and in empirical work with arrays of data. It is consequently worth while to pick out these particular matrices, and introduce their names and properties in this section.

Definition. A matrix all of whose elements are zero is called a *null or zero matrix*, and is denoted by $\mathbf{0}$.

Note that a null matrix does not need to be square. If its order is not apparent from the context, it should be explicitly quoted when the matrix is introduced into any analysis. If an equation in matrices is written with zero on the right-hand side, e.g. $\mathbf{A}=\mathbf{0}$, then the 'zero' is a null matrix of appropriate order.

Definition. A square matrix having unit elements on the principal diagonal and zero elements elsewhere is called an *identity or unit* matrix, and is denoted by

$$\mathbf{I} = \begin{bmatrix} 1 & 0 & 0 & \dots\dots & 0 \\ 0 & 1 & 0 & \dots\dots & 0 \\ \cdot & & & & \\ \cdot & & & & \\ \cdot & & & & \\ \cdot & & & & \\ 0 & 0 & 0 & & 1 \end{bmatrix}$$

There is one such matrix for each order of square matrix. If the order is not clear from the context, it must be specified, usually by writing $\mathbf{I}_n$ for an identity matrix of order n.

The next definition introduces a piece of notation which enables us to describe certain matrices in a very compact manner.

Definition. The Kronecker delta, denoted δ_{ij}, is defined as

$$\left.\begin{aligned} \delta_{ij} &= 1 \quad \text{if } i = j \\ \delta_{ij} &= 0 \quad \text{if } i \neq j \end{aligned}\right\} \quad \text{for } i, j = 1, 2, \dots, n$$

(If you have not previously met the symbol $\neq$, note that it means 'not equal to'.) With the aid of the Kronecker delta notation, an identity matrix can de defined by its typical element:

$$\mathbf{I} = [\delta_{ij}]$$

Another notational convention which is sometimes useful is to write a matrix not in terms of its scalar elements, but as a row of column vectors (or as a column of row vectors). In the present case, using a row of column vectors, we may write

$$\mathbf{I}_n = [\mathbf{e}_1 \ \mathbf{e}_2 \ \dots \ \mathbf{e}_n]$$

where the $\mathbf{e}_j$ are the unit (column) vectors.

Definition. For any scalar λ, a square matrix $\lambda\mathbf{I}$ is called a *scalar matrix.*

This term is not a very happy one since a scalar matrix is not a scalar, unless $\mathbf{I}$ happens to be of order 1 (in which case $\mathbf{I}$ and $\lambda\mathbf{I}$ can be regarded either as a matrix or as a scalar). Of course, there is a different scalar matrix (for given λ) for each identity matrix of different order. The next concept is a generalization of a scalar matrix.

Definition. For scalars λ_i ($i = 1, 2, \dots, n$), a square matrix whose typical element is $[\lambda_i \delta_{ij}]$ is called a *diagonal matrix.*

Note here that the λ_i may vary with i; this distinguishes it from a scalar matrix.

Examples:

(1) $\begin{bmatrix} 3 & 0 \\ 0 & 3 \end{bmatrix} = 3\begin{bmatrix} 1 & 0 \\ 0 & 1 \end{bmatrix}$ is a scalar matrix

(2) $\begin{bmatrix} 4 & 0 & 0 \\ 0 & -2 & 0 \\ 0 & 0 & 3 \end{bmatrix}$ is a diagonal matrix

Definition. A square matrix having zero value for each of its elements below the principal diagonal, and having at least one non-zero element above the principal diagonal (as well as any non-zero elements *on* the diagonal) is called an *upper triangular matrix.*

Equivalently, we may define a *lower triangular matrix* in which the roles of the regions above and below the principal diagonal are reversed.

Examples:

(3) $\begin{bmatrix} 2 & 3 & 1 \\ 0 & 4 & -2 \\ 0 & 0 & 1 \end{bmatrix}$ and $\begin{bmatrix} 2 & 0 & 1 \\ 0 & 4 & 0 \\ 0 & 0 & 1 \end{bmatrix}$ are upper triangular matrices.

(4) $\begin{bmatrix} 2 & 0 & 0 \\ -1 & 4 & 0 \\ 2 & 3 & 1 \end{bmatrix}$ is a lower triangular matrix.

(5) $\begin{bmatrix} 2 & 0 & 0 \\ 0 & 4 & 0 \\ 0 & 0 & 1 \end{bmatrix}$ is classified as a diagonal matrix rather than as a triangular matrix because it has no non-zero elements except on the principal diagonal.

Definition. The *trace* of a square matrix **A** of order *n*, is the sum of the elements on the principal diagonal, and is denoted

$$\text{tr}\,\mathbf{A} = \sum_{i=1}^{n} a_{ii}$$

Note that the trace is a scalar.

Examples:

(6) A special class of matrices which is frequently useful in social science applications is the class of square matrices having each element either zero or unity. Of course the identity matrix is a member of this class. But matrices with other patterns of unit entries also occur. One application (which leads to a haphazard pattern of such entries) is a *connectivity matrix* which shows the connections between a set of entities; perhaps the simplest way to explain this is by means of an example. Suppose that there is a set of n people, and that the connection of interest is the giving of a birthday present by one person in the set to another. Then we might represent all these relationships or connections by a connectivity matrix $\mathbf{C}$ in which the elements c_{ij} are defined

$c_{ij} = 1$ if i^{th} person gives a present to the jth person
$c_{ij} = 0$ if i^{th} person does not give a present to the jth person

This is satisfactory provided $i \neq j$. Where $i = j$ (i.e. for elements on the principal diagonal), we need to adopt some convention, probably to define all $c_{ii} = 0$, but without assigning any connectivity interpretation to these elements. Such a matrix is a neat way of storing a great deal of information. As a numerical example, suppose that persons 1, 2 and 3 each give presents to person 4; 1 and 4 give presents to 3; 2 is not very popular and does not get any presents at all; and 1 receives a present from 4. In this case, the matrix is

$$\mathbf{C} = \begin{bmatrix} 0 & 0 & 1 & 1 \\ 0 & 0 & 0 & 1 \\ 0 & 0 & 0 & 1 \\ 1 & 0 & 1 & 0 \end{bmatrix}$$

In addition to serving as a method of storing information, representation of such an applied situation by a matrix may facilitate analysis of a problem.

(7) A particular type of square matrix with unity and zero entries following a specific pattern is the *permutation matrix*, which has exactly one unity entry in each row and in each column. This can also be thought of as a special type of connectivity matrix, as the following application shows. Suppose that n people have to be assigned to n jobs, each person is able to do any of the

jobs, and the assignment is to be designed so that (in some sense not explored here) the people get the jobs to which they are best suited, when the group of people is considered as a whole. Now define a matrix of variables $\mathbf{X} = [x_{ij}]$ where (for *all* i, j including the cases where $i = j$)

$$x_{ij} = 1 \text{ if } i^{\text{th}} \text{ person gets } j^{\text{th}} \text{ job}$$
$$x_{ij} = 0 \text{ if } i^{\text{th}} \text{ person does not get } j^{\text{th}} \text{ job}$$

(Note that this time the elements on the principal diagonal do not need special attention.) Since the ith person must do a job but cannot do more than one job at once, then any feasible solution for the x_{ij} variables must satisfy the equation

$$\sum_{j=1}^{n} x_{ij} = 1$$

and there is one such equation for each of the n people. Similarly the jth job must be done, but can be assigned to only one person, and this leads to the requirement that

$$\sum_{i=1}^{n} x_{ij} = 1$$

This gives a further set of n equations, one relating to each job. Thus, mathematically speaking, the task is to choose a set of values (each unity or zero) for the x_{ij} variables, so as to satisfy these two sets of equations (and also to give a 'desirable' assignment). Clearly, a considerable number of alternative assignments could be considered; because of the $2n$ constraint equations, each one of these assignments comprises a permutation matrix. As an example, suppose there are four people and four jobs. One feasible assignment is

$$\mathbf{X} = \begin{bmatrix} 0 & 1 & 0 & 0 \\ 0 & 0 & 1 & 0 \\ 1 & 0 & 0 & 0 \\ 0 & 0 & 0 & 1 \end{bmatrix}$$

This permutation matrix shows that the first person does job 2, the second person job 3, the third person job 1 and the fourth person job 4. (The same concepts may be used in other assignment problems, e.g. of factories to locations; and permutation matrices also occur in other types of problem.)

3.5 *Exercises*

1. Prove that tr $\mathbf{A}$+tr $\mathbf{B}$ = tr ($\mathbf{A}+\mathbf{B}$) if $\mathbf{A}$ and $\mathbf{B}$ are square matrices of order n.
*2. If $\mathbf{D}$ denotes a diagonal matrix of order n based on a set of scalars λ_i such that each $\lambda_i \geqq 2$, show that $\mathbf{D} \geqq \mathbf{I}_n$. What can be said about tr $\mathbf{D}$?
*3. If $\mathbf{A}$ is a diagonal matrix of order n with $\lambda_i = i \quad (i = 1, \ldots, n)$ and if $\mathbf{B}_j \ (j = 1, \ldots, n)$ are diagonal matrices of order n with

$$\begin{aligned} \lambda_i &= 0 \qquad i < j \\ \lambda_i &= 1 \qquad i \geqq j \end{aligned}$$

show that
$$\mathbf{A} = \sum_{j=1}^{n} \mathbf{B}_j$$

3.6 *Matrix multiplication – definition*

In defining addition and subtraction of matrices, we took the obvious path, namely that of associating corresponding elements. When it comes to the question of defining a multiplication operation, we could if we wished produce a parallel definition. Although we shall not adopt this approach, it is worth while to examine briefly what it would look like. For such a definition of multiplication, our matrices $\mathbf{A}$ and $\mathbf{B}$ need to be of the same order, say $m \times n$. We could then speak of multiplying the matrices by multiplying each element of $\mathbf{A}$ into the corresponding element of $\mathbf{B}$ to obtain the corresponding element of the product matrix $\mathbf{C}$; in other words

$$c_{ij} = a_{ij} b_{ij} \quad (\text{all } i, j)$$

Now in fact, when processing tables or arrays of data, we do sometimes wish to carry out precisely this kind of operation. For example, the b_{ij} might represent the numbers of people living in grid areas into which the entire area of a country had been divided. For such a grid approach, it would be natural to represent the results in a rectangular array, in which one subscript referred to the latitude and the other to the longitude of each grid area. Now suppose that we wished to estimate car ownership in each grid area by classifying each area according to socio-economic characteristics, and then applying car ownership coefficients (so many cars per 1000 of population) where the coefficients vary according to the socio-economic classification. The complete set of such coefficients, ready for application, might be stored as a rectangular array, the a_{ij}. These could then be multiplied into the b_{ij} (in accordance with our present concept of the multiplication operation) to produce the matrix **C** whose elements c_{ij} represent estimated numbers of cars owned by persons living in each grid area.

While this operation is both respectable and useful in some applications, it is not the one which is conventionally described as matrix multiplication; accordingly we would need another name for it if we wished to make frequent reference to it. Let us now turn to the customary definition of matrix multiplication, which has little to do with the concept of a matrix as an array of data. Instead, it is chosen because it is very useful in the manipulation and solution of systems of simultaneous equations; this use will be illustrated later. The basic idea is to multiply the elements in rows of the first matrix into the elements of columns in the second matrix. Thus the operation is defined only when the first matrix has the same number of columns as the second matrix has rows (thereby ensuring that a row of the first matrix has the same number of elements as a column of the second matrix). A formal definition can now be given:

Definition. Given a matrix **A** of order $m \times n$ and a matrix **B** of order $n \times p$, the operation of multiplying **A** into **B** yields a matrix **C** of order $m \times p$, whose elements are computed from the elements of **A** and **B** according to the formula

$$c_{ij} = \sum_{k=1}^{n} a_{ik} b_{kj} \quad i = 1, \ldots, m; \quad j = 1, \ldots, p$$

In terms of the short-hand notation for each matrix, we write $\mathbf{C} = \mathbf{AB}$ where **A** is called the premultiplying matrix and **B** is called the postmultiplying matrix; as we shall see in detail later, the order in which the matrices are listed does matter. As already noted, the multiplication operation is defined only if the number of columns in the premultiplier equals the number of rows in the postmultiplier. If this condition is satisfied, **A** and **B** are said to be *conformable* for multiplication giving the product **AB**. Note that **C** has the same number of rows as **A** and the same number of columns as **B**, and that there are no restrictions on either of these numbers.

Example:
The significance of this definition of matrix multiplication can be conveniently explored in terms of an example. Let

$$\mathbf{A} = \begin{bmatrix} 2 & 7 \\ 1 & 3 \end{bmatrix} \quad \text{and} \quad \mathbf{B} = \begin{bmatrix} 3 & 2 & 0 \\ 4 & 1 & -2 \end{bmatrix}$$

Note first that the matrices are conformable for multiplication to yield $\mathbf{C} = \mathbf{AB}$, since $\mathbf{A}$ has two columns and $\mathbf{B}$ has two rows. In the computation of the value of c_{ij}, the summation

$$\sum_{k=1}^{n} a_{ik} b_{kj}$$

(which is often called an inner product) involves the sum of n products, one for each value of k; in the present case, $n = 2$. The subscripts i and j define, respectively, a row from $\mathbf{A}$ and a column from $\mathbf{B}$. Consider $i = 1$ and $j = 3$. Then in the present example

$$[a_{11} \quad a_{12}] = [2 \quad 7]$$

and

$$\begin{bmatrix} b_{13} \\ b_{23} \end{bmatrix} = \begin{bmatrix} 0 \\ -2 \end{bmatrix}$$

Thus we compute c_{13} by applying the formula and letting k range over the values 1 and 2:

$$c_{13} = (2 \times 0) + (7 \times (-2)) = -14$$

In words, we multiply the elements in the first row of $\mathbf{A}$ into those in the third column of $\mathbf{B}$ to obtain the element located in the first row and third column of $\mathbf{C}$. Notice also that each row of $\mathbf{A}$ is used several times; for example the first row of $\mathbf{A}$ is multiplied into the second column of $\mathbf{B}$ to give c_{12}, and into the first column of $\mathbf{B}$ to give c_{11}. Similarly each column of $\mathbf{B}$ is used several times. The complete calculation may now be set out:

$$\mathbf{C} = \mathbf{AB} = \begin{bmatrix} (2\times 3)+(7\times 4) & (2\times 2)+(7\times 1) & (2\times 0)+(7\times(-2)) \\ (1\times 3)+(3\times 4) & (1\times 2)+(3\times 1) & (1\times 0)+(3\times(-2)) \end{bmatrix}$$

$$= \begin{bmatrix} 34 & 11 & -14 \\ 15 & 5 & -6 \end{bmatrix}$$

With this definition of matrix multiplication in mind, we can now look at the multiplication of one vector into another, a task postponed from the previous chapter. As already noted (in section 3.1), a vector is a special case of a matrix, having either one row and many columns (in the case of a row vector), or many rows and one column (in the case of a column vector). Suppose matrices **A** of order $1 \times n$ and **B** of order $n \times 1$, i.e. **A** and **B** are a row and a column vector, respectively. The matrix product **AB** is defined since **A** and **B** are conformable for this multiplication. We can adopt this operation as our definition of *vector multiplication*, thus ensuring consistency between this operation and the more general case of matrix multiplication. Let us denote the row vector now by **a**′ and the column vector by **b** (to conform with the notation introduced in Chapter 2), and let

$$\mathbf{a}' = [a_1 \quad a_2 \ldots . a_n] \qquad \mathbf{b} = \begin{bmatrix} b_1 \\ b_2 \\ . \\ . \\ . \\ . \\ b_n \end{bmatrix}$$

Then $$\mathbf{a}'\mathbf{b} = \sum_{i=1}^{n} a_i b_i \text{ which is (of course) a scalar.}$$

Note (i) that this operation is defined only for the multiplication of a row vector into a column vector (and it is for this reason that we use the prime notation to distinguish between the two vector types); this restriction is made solely to ensure consistency between vector multiplication and matrix multiplication.

(ii) that this definition of vector multiplication is that which is suggested by the operation discussed in exercise 5 of section 2.3.

Example:

If $$\mathbf{a}' = [3 \quad 4 \quad -7] \quad \text{and} \quad \mathbf{b}' = [2 \quad -1 \quad 6]$$
$$\mathbf{a}'\mathbf{b} = (3 \times 2) + (4 \times (-1)) + ((-7) \times 6) = -40$$

3.7 *Exercises*

Although matrix multiplication has now been defined, its properties remain to be explored in the next section. Before coming to that, a few exercises are included here just to help you grasp (and remember) the definition.

*1. If $\mathbf{a}' = [2 \quad 7 \quad 8]$ and $\mathbf{b} = \begin{bmatrix} 3 \\ -1 \\ 1 \end{bmatrix}$, is the vector product $\mathbf{a'b}$ defined? If so, calculate the result.

2. If $\mathbf{A} = \begin{bmatrix} 2 & 1 \\ 0 & 2 \end{bmatrix}$, $\mathbf{B} = \begin{bmatrix} 1 & -1 & 0 \\ 2 & 0 & 1 \end{bmatrix}$ and $\mathbf{C} = \begin{bmatrix} 1 & 4 \\ 0 & 2 \\ 3 & -1 \end{bmatrix}$

check in each of the following cases whether the matrix product is defined, and calculate it where possible:

*(a) **AB**
(b) **BC**
*(c) **AC**

3.8 *Some properties of matrix multiplication*

The definition of matrix multiplication turns out to be much more complicated than that for the operation of multiplying one scalar into another. Thus it is hardly surprising that matrix multiplication has properties which sometimes differ from those of scalar multiplication.

Scalar multiplication is commutative; for example $2 \times 7 = 7 \times 2$. But this property does not hold *in general* for matrix multiplication. Suppose **A** is a matrix of order $m \times n$ and **B** a matrix of order $n \times p$, where we assume $m \neq p$. Then **AB** is defined and this product matrix is of order $m \times p$. But the operation is not even defined when we attempt to use **B** as the premultiplying matrix and **A** as the postmultiplier, since **B** has p columns and **A** has m rows. Of course if $m = p$, or in other words, if **A** is of order $m \times n$ and **B** is of order $n \times m$ (with $m \neq n$), then both **AB** and **BA** are defined. But the two product matrices will be different, since **AB** is of order $m \times m$ and **BA** is of order $n \times n$. If **A** and **B** are square of order n, then both **AB** and **BA** are defined, and are also square matrices of order n; in general they will differ, but in any particular case, where **AB** = **BA**, the matrices **A** and **B** are said to commute. Because multiplication is not in general commutative, we can not speak of multiplying **A** and **B**. Instead we must say we will premultiply **A** into **B** (giving **AB**) or postmultiply **B** by **A** (giving **BA**).

Examples:

(1) If $\mathbf{A} = \begin{bmatrix} 2 & 3 \\ 1 & 0 \end{bmatrix}$ and $\mathbf{B} = \begin{bmatrix} 1 & 2 & 3 \\ 0 & 2 & 1 \end{bmatrix}$

$$\mathbf{AB} = \begin{bmatrix} 2 & 10 & 9 \\ 1 & 2 & 3 \end{bmatrix} \quad \text{but } \mathbf{BA} \text{ is not defined}$$

(2) If $\mathbf{A} = \begin{bmatrix} 2 & 3 \\ 1 & 0 \\ 4 & 1 \end{bmatrix}$ and $\mathbf{B} = \begin{bmatrix} 1 & 2 & 3 \\ 0 & 2 & 1 \end{bmatrix}$

$$\mathbf{AB} = \begin{bmatrix} 2 & 10 & 9 \\ 1 & 2 & 3 \\ 4 & 10 & 13 \end{bmatrix} \quad \text{and} \quad \mathbf{BA} = \begin{bmatrix} 16 & 6 \\ 6 & 1 \end{bmatrix}$$

But notice that **AB** and **BA** are not of the same order; *a fortiori* they cannot be said to be equal.

(3) If $\mathbf{A} = \begin{bmatrix} 2 & 0 \\ 0 & 4 \end{bmatrix}$ $\quad \mathbf{B} = \begin{bmatrix} 1 & 1 \\ 2 & 3 \end{bmatrix}$ and $\mathbf{C} = \begin{bmatrix} 1 & 0 \\ 0 & -3 \end{bmatrix}$

then $\mathbf{AB} = \begin{bmatrix} 2 & 2 \\ 8 & 12 \end{bmatrix}$ $\quad \mathbf{BA} = \begin{bmatrix} 2 & 4 \\ 4 & 12 \end{bmatrix}$

Thus $\mathbf{AB} \neq \mathbf{BA}$.

$$\mathbf{A} = \begin{bmatrix} 2 & 0 \\ 0 & -12 \end{bmatrix} \quad \mathbf{CA} = \begin{bmatrix} 2 & 0 \\ 0 & -12 \end{bmatrix}$$

Thus $\mathbf{AC} = \mathbf{CA}$, and matrices **A** and **C** commute.

Matrix multiplication does have the usual associative and distributive properties:

$$(\mathbf{AB})\mathbf{C} = \mathbf{A}(\mathbf{BC}) = \mathbf{ABC}$$
$$\mathbf{A}(\mathbf{B}+\mathbf{C}) = \mathbf{AB}+\mathbf{AC}$$

where **A**, **B** and **C** are matrices of such orders as to make these operations defined. These properties can be proved by examining the typical elements of the product matrices. Consider here just the distributive property:

The $(i, j)^{\text{th}}$ element of **AB** is $\sum_{k=1}^{n} a_{ik} b_{kj}$,

that of **AC** is $\sum_{k=1}^{n} a_{ik} c_{kj}$

While that of $\mathbf{A}(\mathbf{B}+\mathbf{C})$ is $\sum_{k=1}^{n} a_{ik}(b_{kj}+c_{kj})$.

All this follows immediately from the definition of matrix multiplication.

$$\text{Now} \sum_{k=1}^{n} a_{ik}(b_{kj}+c_{kj}) = \sum_{k=1}^{n} a_{ik}b_{kj} + \sum_{k=1}^{n} a_{ik}c_{kj}$$

Thus since the typical elements are equal, the entire matrices must be equal. In other words, $\mathbf{A}(\mathbf{B}+\mathbf{C}) = \mathbf{AB}+\mathbf{AC}$.

Example:

(4) If **A**, **B** and **C** are as defined in example (3), then

$$(\mathbf{AB})\mathbf{C} = \begin{bmatrix} 2 & 2 \\ 8 & 12 \end{bmatrix}\begin{bmatrix} 1 & 0 \\ 0 & -3 \end{bmatrix} = \begin{bmatrix} 2 & -6 \\ 8 & -36 \end{bmatrix}$$

$$\text{Also} \quad \mathbf{BC} = \begin{bmatrix} 1 & 1 \\ 2 & 3 \end{bmatrix}\begin{bmatrix} 1 & 0 \\ 0 & -3 \end{bmatrix} = \begin{bmatrix} 1 & -3 \\ 2 & -9 \end{bmatrix}$$

$$\therefore \quad \mathbf{A}(\mathbf{BC}) = \begin{bmatrix} 2 & 0 \\ 0 & 4 \end{bmatrix}\begin{bmatrix} 1 & -3 \\ 2 & -9 \end{bmatrix} = \begin{bmatrix} 2 & -6 \\ 8 & -36 \end{bmatrix}$$

$$\text{Thus} \quad (\mathbf{AB})\mathbf{C} = \mathbf{A}(\mathbf{BC})$$

$$\text{Also} \quad \mathbf{B}+\mathbf{C} = \begin{bmatrix} 2 & 1 \\ 2 & 0 \end{bmatrix}$$

$$\therefore \quad \mathbf{A}(\mathbf{B}+\mathbf{C}) = \begin{bmatrix} 2 & 0 \\ 0 & 4 \end{bmatrix}\begin{bmatrix} 2 & 1 \\ 2 & 0 \end{bmatrix} = \begin{bmatrix} 4 & 2 \\ 8 & 0 \end{bmatrix}$$

$$\text{While} \quad \mathbf{AB}+\mathbf{AC} = \begin{bmatrix} 2 & 2 \\ 8 & 12 \end{bmatrix}\begin{bmatrix} 2 & 0 \\ 0 & -12 \end{bmatrix} = \begin{bmatrix} 4 & 2 \\ 8 & 0 \end{bmatrix}$$

$$\text{Thus} \quad \mathbf{A}(\mathbf{B}+\mathbf{C}) = \mathbf{AB}+\mathbf{AC}.$$

We are now in a position to look at the multiplicative properties of some of the special types of matrix which were introduced in section 3.4. Consider first the null matrix. In the algebra of scalars, $0 \times \lambda = \lambda \times 0 = 0$. The null matrix has properties which are similar to this. Suppose **A** is of order $m \times n$. Then it follows directly from the definition of matrix multiplication that

$$\mathbf{A0} = \mathbf{0} \tag{3–1}$$

where on the left-hand side the null matrix must have n rows (in order to ensure that multiplication is defined). It may have p columns where p is any integer greater than or equal to 1. The null matrix on the right-hand side is then of order $m \times p$.

Similarly

$$\mathbf{0A} = \mathbf{0} \tag{3–2}$$

where this time the two null matrices must be of order $p \times m$ and $p \times n$ respectively. If $\mathbf{A}$ is a square matrix of order n, then

$$\mathbf{A0} = \mathbf{0A} = \mathbf{0}$$

where all null matrices are also square of order n. (Notice that (3–1) and (3–2) cannot be combined into a single matrix equation, since the former implies m rows for each of the matrices, and the latter implies p rows where – in general – $p \neq m$.)

What might be termed 'cancellation' of matrices (by analogy with scalar algebra) is not permitted by the definitions. If λ and ϕ are scalars, and if $\lambda\phi = 0$, then this means that $\lambda = 0$, or $\phi = 0$, or both equal zero. But the matrix equation $\mathbf{AB} = \mathbf{0}$ does not necessarily imply that either $\mathbf{A} = \mathbf{0}$ or $\mathbf{B} = \mathbf{0}$; one of the following examples shows that the product of two matrices can be a null matrix even though neither of the matrices being multiplied is a null matrix.

Examples:

(5) $$\begin{bmatrix} 1 & -7 \\ 3 & 2 \end{bmatrix} \begin{bmatrix} 0 & 0 \\ 0 & 0 \end{bmatrix} = \begin{bmatrix} 0 & 0 \\ 0 & 0 \end{bmatrix}$$

(6) $$\begin{bmatrix} 0 & 0 \\ 2 & -3 \end{bmatrix} \begin{bmatrix} 3 & 0 \\ 2 & 0 \end{bmatrix} = \begin{bmatrix} 0 & 0 \\ 0 & 0 \end{bmatrix}$$

The multiplication of the identity matrix into a general matrix $\mathbf{A}$ of order $m \times n$ has a very simple effect. We can use $\mathbf{I}_m$ (an identity matrix of order m) as a premultiplying matrix:

$$\mathbf{I}_m\mathbf{A} = \mathbf{A}$$

as may be seen by direct application of the definition of multiplication. For instance, consider

$$\begin{bmatrix} 1 & 0 \\ 0 & 1 \end{bmatrix} \begin{bmatrix} a_{11} & a_{12} & a_{13} \\ a_{21} & a_{22} & a_{23} \end{bmatrix} = \begin{bmatrix} a_{11} & a_{12} & a_{13} \\ a_{21} & a_{22} & a_{23} \end{bmatrix}$$

Similarly we may postmultiply $\mathbf{A}$ by $\mathbf{I}_n$ giving

$$\mathbf{AI}_n = \mathbf{A}$$

If **A** is square of order n, and we take an identity matrix of order n,

$$\mathbf{AI} = \mathbf{IA} = \mathbf{A}$$

Thus **I** and **A** commute. If we set $\mathbf{A} = \mathbf{I}$, then the equation becomes

$$\mathbf{II} = \mathbf{I}$$

It seems reasonable to adopt the notational convention $\mathbf{II} = \mathbf{I}^2$. Similarly $\mathbf{I}^3 = \mathbf{I}$, $\mathbf{I}^4 = \mathbf{I}$ and so on, provided all matrices are of the same order.

In summary, multiplication of a matrix by the identity matrix (of appropriate order) leaves the first matrix unaltered – which is why the identity matrix is so called. (This may raise doubts as to whether the identity matrix is ever useful; the answer is that it is, as will be seen later.)

Finally, just as scalar and diagonal matrices are generalizations of the identity matrix, so are the effects of multiplying by these matrices generalizations of the effect of multiplying by an identity matrix. If a matrix is premultiplied or postmultiplied by a scalar matrix, all its elements are multiplied by the value of the scalar λ. If a diagonal matrix is used as a premultiplying matrix, the rows of the other matrix have their elements multiplied by the corresponding λ_i; in postmultiplication by a diagonal matrix, the elements in *columns* are multiplied by the λ_i. The following examples illustrate all the effects.

Examples:

(7) $$\begin{bmatrix} 3 & 0 \\ 0 & 3 \end{bmatrix} \begin{bmatrix} 4 & 2 \\ -1 & 0 \end{bmatrix} = \begin{bmatrix} 12 & 6 \\ -3 & 0 \end{bmatrix}$$

(8) $$\begin{bmatrix} 4 & 2 \\ -1 & 0 \end{bmatrix} \begin{bmatrix} 3 & 0 \\ 0 & 3 \end{bmatrix} = \begin{bmatrix} 12 & 6 \\ -3 & 0 \end{bmatrix}$$

(9) $$\begin{bmatrix} \lambda_1 & 0 \\ 0 & \lambda_2 \end{bmatrix} \begin{bmatrix} 2 & -7 \\ 1 & 3 \end{bmatrix} = \begin{bmatrix} 2\lambda_1 & -7\lambda_1 \\ \lambda_2 & 3\lambda_2 \end{bmatrix}$$

(10) $$\begin{bmatrix} 2 & -7 \\ 1 & 3 \end{bmatrix} \begin{bmatrix} \lambda_1 & 0 \\ 0 & \lambda_2 \end{bmatrix} = \begin{bmatrix} 2\lambda_1 & -7\lambda_2 \\ \lambda_1 & 3\lambda_2 \end{bmatrix}$$

3.9 *Exercises*

1. In each of the following cases, consider whether the product matrices **AB** *and* **BA** are defined, and evaluate them whenever they are:

(a) $\mathbf{A} = \begin{bmatrix} 2 & 3 \\ 5 & -1 \end{bmatrix}$ $\quad\mathbf{B} = \begin{bmatrix} -2 & 7 \\ 4 & 0 \end{bmatrix}$

(b) $\mathbf{A} = \begin{bmatrix} 3 & 2 & 6 & 8 \\ 4 & -1 & 3 & 7 \end{bmatrix}$ $\quad\mathbf{B} = \begin{bmatrix} 2 & 3 \\ -1 & 0 \\ 0 & 1 \end{bmatrix}$

*(c) $\mathbf{A} = [2 \quad 7 \quad 4 \quad -1]$ $\quad\mathbf{B} = \begin{bmatrix} 3 \\ 7 \\ 2 \end{bmatrix}$

(d) $\mathbf{A} = [2 \quad 3]$ $\quad\mathbf{B} = \begin{bmatrix} 4 \\ 1 \end{bmatrix}$

*2. If $\mathbf{A} = \begin{bmatrix} 1 & 2 \\ 0 & 0 \end{bmatrix}$ and if $\mathbf{B} = \begin{bmatrix} \alpha & \beta \\ 1 & 1 \end{bmatrix}$

what conditions must be placed on α and β to ensure that the product **AB** is a null matrix of appropriate order?

3. What properties must matrices **A** and **B** possess to ensure that

$$(\mathbf{A}+\mathbf{B})^2 = \mathbf{A}^2+2\mathbf{AB}+\mathbf{B}^2?$$

(Note: $\mathbf{A}^2$ means simply **AA**, with similar interpretation for $\mathbf{B}^2$ and for $(\mathbf{A}+\mathbf{B})^2$.)

4. If the diagonal matrices $\mathbf{A} = [\lambda_i\, \delta_{ij}]$ and $\mathbf{B} = [\phi_i\, \delta_{ij}]$ are both of order n, show that they commute.

3.10 *Summation notation again*

Let us now develop an extension of the use of the summation notation first introduced in section 1.4; this extension will prepare the way for the discussion (in the next section) of matrix representation of simultaneous linear equations.

Begin by considering a matrix of order $m \times n$, having as its typical element the product $f_i g_j$, where f_i and g_j are scalars. Written out in full, the matrix is

$$\begin{bmatrix} f_1g_1 & f_1g_2 & \cdot & \cdot & \cdot & \cdot & f_1g_n \\ f_2g_1 & f_2g_2 & \cdot & \cdot & \cdot & \cdot & f_2g_n \\ \cdot & \cdot & & & & & \cdot \\ \cdot & \cdot & & & & & \cdot \\ \cdot & \cdot & & & & & \cdot \\ \cdot & \cdot & & & & & \cdot \\ f_mg_1 & f_mg_2 & \cdot & \cdot & \cdot & \cdot & f_mg_n \end{bmatrix}$$

Now suppose that we want to find an expression to represent the sum of all the $m \times n$ elements in the matrix; this sum is, of course, a single scalar. One way of tackling this is first to add up all the elements in the ith row to give a row sum which may be written as

$$\sum_{j=1}^{n} f_i g_j = f_i \sum_{j=1}^{n} g_j$$

(Note that there are alternative expressions open to us here: since f_i is a constant with respect to the summation, which is over the various values of j, then it is immaterial whether the f_i is written before or after the Σ.) For the next step, we may add up all the row sums, to get the grand total of all the elements, which total can be written in the following equivalent ways:

$$\text{(3–3)} \qquad \sum_{i=1}^{m} \sum_{j=1}^{n} f_i g_j = \sum_{i=1}^{m} f_i \sum_{j=1}^{n} g_j = \sum_{i=1}^{m} \sum_{j=1}^{n} g_j f_i$$

(The first two forms follow from the addition applied to the previous expressions; the third form is clearly the same as the first because $f_i g_j = g_j f_i$, since we are simply multiplying scalars.)

Now an alternative way of adding up all the elements in the matrix is to start by adding all the elements in the jth column to give a column sum written as

$$\sum_{i=1}^{m} f_i g_j = g_j \sum_{i=1}^{m} f_i$$

(The factor g_j may be put in front of the Σ since it is a constant with respect to this summation.) Now add up all the column sums to give the grand total, which may be written alternatively as

$$\text{(3–4)} \qquad \sum_{j=1}^{n} \sum_{i=1}^{m} f_i g_j = \sum_{j=1}^{n} g_j \sum_{i=1}^{m} f_i$$

Clearly the grand total must have the same value whichever method is used to effect the addition. Hence, by comparing equations (3–3) with (3–4), we have

$$\text{(3–5)} \quad \sum_{i=1}^{m} f_i \sum_{j=1}^{n} g_j = \sum_{i=1}^{m} \sum_{j=1}^{n} f_i g_j = \sum_{j=1}^{n} \sum_{i=1}^{m} f_i g_j = \sum_{j=1}^{n} g_j \sum_{i=1}^{m} f_i$$

In other words, when *double summations* are carried out (i.e. where the summation is with respect to *two* subscripts) the order in which the summation is carried out is immaterial; in the final summation expression, either Σ may be written first, as in the second and third expressions in (3–5); and if the summation signs are 'separated', as in the first and fourth expressions in (3–5), then again the order is immaterial, except (of course) that a factor may not be transferred to a position before a Σ unless it is a constant with respect to that summation, e.g. g_j which is brought into a position before $\sum_{i=1}^{m}$ in the fourth expression in (3–5).

Similar results apply when double-subscripted variables are in use, but particular care must be taken when rearranging the order of the summation signs. As an example, consider the expression

$$\sum_{j=1}^{n} a_{ij} \sum_{k=1}^{p} c_{jk} y_k \tag{3–6}$$

The coefficient a_{ij} is a constant with respect to the summation over k, and hence can appear after this summation as an alternative. Also y_k is a constant with respect to the summation over j and hence can appear equivalently before or after that summation sign; on the other hand it varies with k and since it appears in (3–6) after the summation with respect to k, it must always do so in any alternative expression. These remarks show that the following expressions are equivalent to (3–6):

$$\sum_{k=1}^{p} \sum_{j=1}^{n} a_{ij} c_{jk} y_k = \sum_{k=1}^{p} y_k \sum_{j=1}^{n} a_{ij} c_{jk} \tag{3–7}$$

Examples:

(1)

If $\mathbf{A} = \begin{bmatrix} 1 & 2 & 3 & 4 \\ 5 & 6 & 7 & 8 \\ 1 & 0 & 0 & 0 \\ 0 & 1 & 0 & 0 \end{bmatrix}$

then
$$\begin{aligned} \sum_{i=1}^{2} \sum_{j=2}^{3} a_{ij} &= \sum_{i=1}^{2} (a_{i2}+a_{i3}) \\ &= (a_{12}+a_{13})+(a_{22}+a_{23}) \\ &= (2+3)+(6+7) = 18 \end{aligned}$$

(2) For the same matrix $\mathbf{A}$, and noting that a_{ji} is the element in the jth row and i^{th} column,

$$\sum_{i=1}^{2} \sum_{j=2}^{3} a_{ij} a_{ji} = \sum_{i=1}^{2} \left(\sum_{j=2}^{3} a_{ij} a_{ji} \right)$$

$$= \sum_{i=1}^{2} (a_{i2} a_{2i} + a_{i3} a_{3i})$$

$$= (a_{12}a_{21} + a_{13}a_{31}) + (a_{22}a_{22} + a_{23}a_{32})$$

$$= (2 \times 5 + 3 \times 1) + (6 \times 6 + 7 \times 0) = 49$$

(3) For a general matrix **A** and a vector **b**,

$$\sum_{j=1}^{2} \sum_{i=1}^{2} a_{ij} b_i = \sum_{j=1}^{2} (a_{1j}b_1 + a_{2j}b_2)$$

$$= (a_{11}b_1 + a_{21}b_2) + (a_{12}b_1 + a_{22}b_2)$$

Also

$$\sum_{i=1}^{2} b_i \sum_{j=1}^{2} a_{ij} = \sum_{i=1}^{2} b_i(a_{i1} + a_{i2})$$

$$= b_1(a_{11} + a_{12}) + b_2(a_{21} + a_{22})$$

But this is the same set of four terms as before. Thus

$$\sum_{j=1}^{2} \sum_{i=1}^{2} a_{ij}b_i = \sum_{i=1}^{2} b_i \sum_{j=1}^{2} a_{ij}$$

3.11 *Matrix multiplication and simultaneous equations*

In this section we shall see not only that matrices provide a very compact notation for describing systems of simultaneous linear equations, but also that the definition of matrix multiplication is particularly useful in the manipulation of such systems.

Consider a general set of simultaneous equations:

$$\begin{aligned} a_{11}x_1 + a_{12}x_2 + \ldots + a_{1n}x_n &= b_1 \\ a_{21}x_1 + a_{22}x_2 + \ldots + 2_{2n}x_n &= b_2 \\ &\vdots \\ a_{m1}x_1 + a_{m2}x_2 + \ldots + a_{mn}x_n &= b_m \end{aligned} \qquad (3\text{–}8)$$

We have already seen (in section 2.4) that this can be written more compactly using vector notation. This process can be taken further still with matrix notation. If we define $\mathbf{A} = [a_{ij}]$ to be a matrix of order $m \times n$, $\mathbf{x}' = [x_1 \; x_2 \ldots . x_n]$ and $\mathbf{b}' = [b_1 \; b_2 \ldots . b_m]$, then the equations may be written $\mathbf{Ax} = \mathbf{b}$ where $\mathbf{x}$ and $\mathbf{b}$ are (of course) column vectors. Check for yourself, by multiplying $\mathbf{A}$ into $\mathbf{x}$, that this is equivalent to the initial set of equations (3–8).

The manipulation of systems of linear equations will be discussed first in terms of a small example. Consider the set of equations

$$\begin{aligned} 2x_1 + \ x_2 - 2x_3 &= 1 \\ 3x_1 + 6x_2 + \ x_3 &= 8 \end{aligned} \tag{3–9}$$

and suppose that the variables x_1, x_2 and x_3 are themselves expressed in terms of another set of variables (y_1 and y_2) by the equations

$$\begin{aligned} x_1 &= 2y_1 + 7y_2 \\ x_2 &= 5y_1 + \ y_2 \\ x_3 &= 4y_1 - 2y_2 \end{aligned} \tag{3–10}$$

Here we shall not seek to solve the equations – that kind of task is reserved for Chapter 6. Instead we shall simply eliminate the $\mathbf{x}$ variables, to obtain a set of equations in the $\mathbf{y}$ variables alone. First consider how this may be done by employing the operations of scalar algebra: the equations (3–10) may be used to substitute in (3–9) for the $\mathbf{x}$ variables:

$$\begin{aligned} 2(2y_1 + 7y_2) + \ (5y_1 + y_2) - 2(4y_1 - 2y_2) &= 1 \\ 3(2y_1 + 7y_2) + 6(5y_1 + y_2) + \ (4y_1 - 2y_2) &= 8 \end{aligned}$$

After grouping the terms, these equations become

$$\begin{aligned} y_1 + 19y_2 &= 1 \\ 40y_1 + 25y_2 &= 8 \end{aligned} \tag{3–11}$$

and the job is done.

Now, as an alternative, consider how the task could be tackled by using the operations of matrix algebra. First, by defining appropriate matrices, the systems (3–9) and (3–10) may be written out, respectively, as:

$$\begin{bmatrix} 2 & 1 & -2 \\ 3 & 6 & 1 \end{bmatrix} \begin{bmatrix} x_1 \\ x_2 \\ x_3 \end{bmatrix} = \begin{bmatrix} 1 \\ 8 \end{bmatrix} \tag{3–12}$$

$$\begin{bmatrix} x_1 \\ x_2 \\ x_3 \end{bmatrix} = \begin{bmatrix} 2 & 7 \\ 5 & 1 \\ 4 & -2 \end{bmatrix} \begin{bmatrix} y_1 \\ y_2 \end{bmatrix} \tag{3–13}$$

(Check these derivations for yourself.) Now in matrix algebra, just as in scalar algebra, we may use the substitution operation; here we use the system (3–13) to substitute for the **x**-vector in equation (3–12), leading to the matrix equation

$$\begin{bmatrix} 2 & 1 & -2 \\ 3 & 6 & 1 \end{bmatrix} \begin{bmatrix} 2 & 7 \\ 5 & 1 \\ 4 & -2 \end{bmatrix} \begin{bmatrix} y_1 \\ y_2 \end{bmatrix} = \begin{bmatrix} 1 \\ 8 \end{bmatrix}$$

If the first matrix on the left-hand side is multiplied into the second matrix (and note that the matrices are conformable for this multiplication), then the system may be written more simply:

$$\begin{bmatrix} 1 & 19 \\ 40 & 25 \end{bmatrix} \begin{bmatrix} y_1 \\ y_2 \end{bmatrix} = \begin{bmatrix} 1 \\ 8 \end{bmatrix} \tag{3–14}$$

Again, if the matrices on the left-hand side are multiplied together, this result can be expressed in scalars only; it is, of course, then identical with equations (3–11). This example shows that the definition we have adopted for the operation of multiplying matrices is indeed useful when it comes to handling systems of linear equations. In the present case, the work involved in using matrix operations to go from (3–12) and (3–13) to (3–14) is about as much as that required to use scalar operations to obtain (3–11) from (3–9) and (3–10); but in larger examples, the matrix representation gives a neater and quicker way of carrying out the arithmetic.

With the aid of the double summation notation introduced in the previous section, the type of manipulation just explored in the numerical example can be set out for the general case. Consider again the general system (3–8) and suppose that the variables x_j are expressed in terms of a set of p variables called y_k $(k = 1, 2, \ldots., p)$ by means of the linear equations

$$\begin{aligned} x_1 &= c_{11}y_1 + \ldots. + c_{1p}y_p \\ x_2 &= c_{21}y_1 + \ldots. + c_{2p}y_p \\ &\cdot \\ &\cdot \\ &\cdot \\ &\cdot \\ x_n &= c_{n1}y_1 + \ldots. + c_{np}y_p \end{aligned} \tag{3–15}$$

The jth equation from (3–15) may be written

(3–16) $$x_j = \sum_{k=1}^{p} c_{jk} y_k$$

Similarly the ith equation from (3–8) may be written

$$\sum_{j=1}^{n} a_{ij} x_j = b_i$$

and in this equation substitute for each x_j using equations such as (3–16):

$$\sum_{j=1}^{n} a_{ij} \sum_{k=1}^{p} c_{jk} y_k = b_i$$

Now the left-hand side of this expression was studied in section 3.10 as an illustration of the properties of double summations. In particular, from equations (3–6) and (3–7) we know that the above equation may be rewritten as

(3–17) $$\sum_{k=1}^{p} y_k \sum_{j=1}^{n} a_{ij} c_{jk} = b_i$$

The latter part of the left-hand side, $\sum_{j=1}^{n} a_{ij} c_{jk}$, reminds us of the definition of matrix multiplication. Let us define a matrix **C** of order $n \times p$ having as its elements the coefficients from (3–15). The summation $\sum_{j=1}^{n} a_{ij} c_{jk}$ is now seen to be the ith row of **A** multiplied into the kth column of **C**. This suggests that we define a product matrix **D** of order $m \times p$ by the equivalent expressions

$$\mathbf{D} = \mathbf{AC}$$

or

$$d_{ik} = \sum_{j=1}^{n} a_{ij} c_{jk}$$

Equation (3–17) may now be written

$$\sum_{k=1}^{p} d_{ik} y_k = d_{i1} y_1 + d_{i2} y_2 + \ldots + d_{ip} y_p = b_i$$

and this is clearly the ith equation from the set

$$\mathbf{Dy} = \mathbf{b}$$

as you should check for yourself. These then are alternative ways of expressing the set of equations which involve only the **y** variables. These expressions have been derived by studying the details of the individual equations, and by using double summation notation, to gain familiarity with these underlying details of the matrix representation. But once we have an understanding of what is going on 'inside' the matrix equations (so to speak), the derivation can be done much more quickly by working entirely in matrix terms. The sets of equations (3–8) and (3–15) may be written

$$\mathbf{Ax} = \mathbf{b} \quad \text{and} \quad \mathbf{x} = \mathbf{Cy}$$

where **A** is of order $m \times n$ and **C** is of order $n \times p$. Use the second set to substitute for **x**, giving

$$\mathbf{ACy} = \mathbf{b}$$

Define a matrix of order $m \times p$ as $\mathbf{D} = \mathbf{AC}$ and the matrix equation may then be written simply

$$\mathbf{Dy} = \mathbf{b}$$

Once mastered, the matrix notation and its properties really do save effort!

To end the section, consider one further example. Suppose there are two column vectors of variables denoted **x** and **y**, of order m and n respectively, and that these are inter-related by the following two sets of equations:

$$\begin{aligned} \mathbf{Ax} + \mathbf{y} &= \mathbf{f} \\ -\mathbf{x} + \mathbf{By} &= \mathbf{0} \end{aligned}$$

where **A** is of order $n \times m$, **B** is of order $m \times n$ and **f** is a column vector of order n. (Check for yourself that these orders lead to all the matrix multiplications and additions in these equations being properly defined. What must be the order of the null vector on the right-hand side of the second set?) Suppose that our task is to eliminate the **y** variables, to leave a set of equations involving the **x** variables only. This job is quickly done: premultiply the first matrix equation by **B** (noting that the multiplication is defined):

$$\mathbf{BAx} + \mathbf{By} = \mathbf{Bf}$$

From the second equation, we note that

$$\mathbf{By} = \mathbf{x}$$

and this may be used to substitute in the previous equation for **By**, to give

$$\mathbf{BAx}+\mathbf{x} = \mathbf{Bf}$$

i.e.

$$(\mathbf{BA}+\mathbf{I}_m)\mathbf{x} = \mathbf{Bf}$$

(Again check these operations for yourself; and note the use of the identity matrix, of order m, to give a neater expression.)

3.12 *Exercises*

1. If $\mathbf{A} = \begin{bmatrix} 3 & 4 & -1 \\ 0 & 2 & 1 \\ 1 & 2 & 0 \end{bmatrix}$ calculate

(a) $\sum_{i=1}^{3} \sum_{j=1}^{3} a_{ij}$ *(b) $\sum_{i=1}^{2} \sum_{j=1}^{2} a_{ij}a_{i,j+1}$ (c) $\sum_{i=2}^{3} \sum_{j=2}^{3} ij\,a_{ij}$

2. A chain store with n branches sells the same m homogeneous commodities at each branch. The stock of the ith commodity held at the jth branch is denoted a_{ij} $(i=1, 2, \ldots, m; j=1, 2, \ldots, n)$ and is measured in physical units (e.g. if the first commodity is radios, the stock at one branch might be 3 radios, that at another 5 identical radios, and so forth). What physical meaning (if any) can be given to each of the following sums?

*(a) $\sum_{j=1}^{n} a_{ij}$ (b) $\sum_{i=1}^{m} a_{ij}$ *(c) $\sum_{i=1}^{m} \sum_{j=1}^{n} a_{ij}$

3. Employ a matrix representation of the following two sets of equations to eliminate the **y** variables, leaving a system of equations in the **x** variables only:

$$x_1+\ y_1=4$$
$$x_2+\ y_2=3$$

and

$$3y_1+\ y_2=7$$
$$y_1+2y_2=6$$

*4. Suppose that **x** and **y** are (column) vectors of variables, each of order n and that another column vector of variables **z** is of order p. **A**, **B** and **C** are matrices of coefficients, and are of orders $n \times n$, $p \times n$ and $p \times p$ respectively. There are also two vectors of constants, **g** of order n and **h** of order p. Given the following systems of equations, eliminate the **y** and **z** variables to leave a system of equations in the **x** variables:

$$\mathbf{Ax}+\mathbf{y}=\mathbf{g}$$
$$\mathbf{By}=\mathbf{z}$$
$$\mathbf{Cz}=\mathbf{h}$$

How many equations are there in the system thus obtained?

3.13 *Submatrices and partitioning*

It is often useful to focus attention on a part of a matrix, or on some division of the entire matrix into parts, where each part can be more than merely a single element:

Definition. Given a matrix $\mathbf{A}$ of order $m \times n$, if all but r rows and s columns are struck out, the resulting matrix of order $r \times s$ is called a *submatrix* of $\mathbf{A}$.

Examples:

(1) If the third and fourth rows and the fourth column are struck out from the matrix

$$\mathbf{A}=\begin{bmatrix} 2 & 3 & -1 & 4 \\ 0 & 2 & 7 & -2 \\ 3 & 4 & 6 & 1 \\ 2 & 1 & 0 & 2 \end{bmatrix}$$

there remains a submatrix

$$\begin{bmatrix} 2 & 3 & -1 \\ 0 & 2 & 7 \end{bmatrix}$$

(2) The 4×4 matrix of example (1) can be thought of as being composed of four submatrices, where broken lines denote the partitioning:

$$\mathbf{A}=\left[\begin{array}{ccc:c} 2 & 3 & -1 & 4 \\ 0 & 2 & 7 & -2 \\ \hdashline 3 & 4 & 6 & 1 \\ 2 & 1 & 0 & 2 \end{array}\right]$$

This *partitioning* of matrices may sometimes be useful in giving a vivid picture of the structure of $\mathbf{A}$ (for example, $\mathbf{A}$ might include a submatrix which is a null matrix); it may also be useful in simplifying matrix computations, as will be seen later. Note that any given matrix may (in general) be partitioned in a number of different, alternative ways. Attention will be confined here to schemes of partitioning in which submatrices are built up from *adjacent* rows and *adjacent* columns of the original matrix; in other words we are concerned with (square or rectangular) blocks drawn from that matrix. The usual notation is to describe a submatrix by a bold upper-case letter with two subscripts, following an analogy with the double-subscript notation for (scalar) elements of a matrix.

Examples:

(3) The partitioning in example (2) might be denoted

$$\mathbf{A} = \begin{bmatrix} \mathbf{A}_{11} & \mathbf{A}_{12} \\ \mathbf{A}_{21} & \mathbf{A}_{22} \end{bmatrix}$$

where $\mathbf{A}_{11} = \begin{bmatrix} 2 & 3 & -1 \\ 0 & 2 & 7 \end{bmatrix}$ $\quad \mathbf{A}_{12} = \begin{bmatrix} 4 \\ -2 \end{bmatrix}$

$$\mathbf{A}_{21} = \begin{bmatrix} 3 & 4 & 6 \\ 2 & 1 & 0 \end{bmatrix} \qquad \mathbf{A}_{22} = \begin{bmatrix} 1 \\ 2 \end{bmatrix}$$

(4) The matrix $\mathbf{A} = \begin{bmatrix} a_{11} & a_{12} & 1 \\ 0 & 0 & a_{23} \\ 0 & 0 & a_{33} \end{bmatrix}$

might be written $\mathbf{A} = \begin{bmatrix} \mathbf{A}_{11} & \mathbf{I} \\ \mathbf{0} & \mathbf{A}_{22} \end{bmatrix}$

where $\mathbf{A}_{11} = [a_{11} \quad a_{12}]$, $\quad \mathbf{A}_{22} = \begin{bmatrix} a_{23} \\ a_{24} \end{bmatrix}$

I is a unit matrix of order 1, and **0** a null matrix of order 2×2.

The implications for matrix operations are now studied briefly. As an example, consider two matrices **A** and **B** partitioned each into six submatrices:

$$\mathbf{A} = \begin{bmatrix} \mathbf{A}_{11} & \mathbf{A}_{12} & \mathbf{A}_{13} \\ \mathbf{A}_{21} & \mathbf{A}_{22} & \mathbf{A}_{23} \end{bmatrix} \qquad \mathbf{B} = \begin{bmatrix} \mathbf{B}_{11} & \mathbf{B}_{12} & \mathbf{B}_{13} \\ \mathbf{B}_{21} & \mathbf{B}_{22} & \mathbf{B}_{23} \end{bmatrix}$$

Provided that for each $\mathbf{A}_{ij}$ the corresponding $\mathbf{B}_{ij}$ is of the same order, then clearly

$$\mathbf{A}+\mathbf{B} = \begin{bmatrix} \mathbf{A}_{11}+\mathbf{B}_{11} & \mathbf{A}_{12}+\mathbf{B}_{12} & \mathbf{A}_{13}+\mathbf{B}_{13} \\ \mathbf{A}_{21}+\mathbf{B}_{21} & \mathbf{A}_{22}+\mathbf{B}_{22} & \mathbf{A}_{23}+\mathbf{B}_{23} \end{bmatrix}$$

In general then the rule for addition (or subtraction) of partitioned matrices is simply to add (or subtract) corresponding submatrices, given that the matrices and their partitioning are such that corresponding submatrices are conformable for addition (i.e. have the same order).

In order to study multiplication, consider an example in which a matrix **A** of order 2×3 is premultiplied into a matrix **B** of order 3×2, where

$$\mathbf{A} = [\mathbf{A}_{11} \quad \mathbf{A}_{12}] = \left[\begin{array}{cc|c} a_{11} & a_{12} & a_{13} \\ a_{21} & a_{22} & a_{23} \end{array}\right] \text{ and } \mathbf{B} = \begin{bmatrix} \mathbf{B}_{11} \\ \mathbf{B}_{21} \end{bmatrix} = \left[\begin{array}{cc} b_{11} & b_{12} \\ b_{21} & b_{22} \\ \hline b_{31} & b_{32} \end{array}\right]$$

Notice that the product matrix $\mathbf{C} = \mathbf{AB}$ is defined, and that the *columns* of the premultiplying matrix **A** have been partitioned in the same ways as the *rows* of the postmultiplying matrix **B**, from which it follows that the products $\mathbf{A}_{11}\mathbf{B}_{11}$ and $\mathbf{A}_{12}\mathbf{B}_{21}$ of submatrices are also defined. The general proposition is that provided these submatrix products are defined, the product matrix **C** can be expressed by applying the usual matrix multiplication operation to the submatrices just as if they were scalar elements. In the present case, this yields

$$\mathbf{C} = [\mathbf{A}_{11} \quad \mathbf{A}_{12}] \begin{bmatrix} \mathbf{B}_{11} \\ \mathbf{B}_{21} \end{bmatrix} = [\mathbf{A}_{11}\mathbf{B}_{11} + \mathbf{A}_{12}\mathbf{B}_{21}]$$

In this case, it can be readily verified that this rule gives the correct result:

$$\mathbf{A}_{11}\mathbf{B}_{11} = \begin{bmatrix} a_{11}b_{11} + a_{12}b_{21} & a_{11}b_{12} + a_{12}b_{22} \\ a_{21}b_{11} + a_{22}b_{21} & a_{21}b_{12} + a_{22}b_{22} \end{bmatrix}$$

and $$\mathbf{A}_{12}\mathbf{B}_{21} = \begin{bmatrix} a_{13}b_{31} & a_{13}b_{32} \\ a_{23}b_{31} & a_{23}b_{32} \end{bmatrix}$$

Thus $\mathbf{A}_{11}\mathbf{B}_{11} + \mathbf{A}_{12}\mathbf{B}_{21}$

$$= \begin{bmatrix} a_{11}b_{11} + a_{12}b_{21} + a_{13}b_{31} & a_{11}b_{12} + a_{12}b_{22} + a_{13}b_{32} \\ a_{21}b_{11} + a_{22}b_{21} + a_{23}b_{31} & a_{21}b_{12} + a_{22}b_{22} + a_{23}b_{32} \end{bmatrix}$$

which is, of course, the same result as that obtained by direct premultiplication of **A** into **B**. The way in which this result comes about suggests that a similar mechanism is at work in any such case in which the columns of **A** are partitioned in the same way as the rows of **B**. In fact a general proof can be devised which is based on this; however the proof is not given here. Note that the rows of **A** and the columns of **B** can also be partitioned, and do not need to be partitioned in the same way. Notice also that partitioning which permits the premultiplication of **A** into **B** will not in general permit the premultiplication of **B** into **A** (although it might in special cases). Similarly partitioning suitable for multiplication will not in general be suitable for addition.

Example:

(5) $$\mathbf{A} = [\mathbf{A}_{11} \quad \mathbf{A}_{12}] = \left[\begin{array}{c|c} 3 & 0 \\ 7 & 0 \end{array}\right]$$

$$\mathbf{B} = \begin{bmatrix} \mathbf{B}_{11} & \mathbf{B}_{12} \\ \mathbf{B}_{21} & \mathbf{B}_{22} \end{bmatrix} = \left[\begin{array}{cc|c} 4 & -2 & 6 \\ \hline 3 & 1 & 1 \end{array}\right]$$

$$\mathbf{AB} = [\mathbf{A}_{11}\mathbf{B}_{11} + \mathbf{A}_{12}\mathbf{B}_{21} \quad \mathbf{A}_{11}\mathbf{B}_{12} + \mathbf{A}_{12}\mathbf{B}_{22}]$$

But $\mathbf{A}_{12}$ is a null matrix. $\therefore$ $\mathbf{A}_{12}\mathbf{B}_{21}$ and $\mathbf{A}_{12}\mathbf{B}_{22}$ are null matrices.

Thus $$\mathbf{AB} = [\mathbf{A}_{11}\mathbf{B}_{11} \quad \mathbf{A}_{11}\mathbf{B}_{12}]$$
$$= \begin{bmatrix} 12 & -6 & 18 \\ 28 & -14 & 42 \end{bmatrix}$$

Compare this for yourself with the operation of premultiplying **A** into **B**.

3.14 *Transposition; symmetric and skew-symmetric matrices*

We now introduce *transposition*, a matrix operation which has no counterpart in scalar algebra. A matrix of order $m \times n$ can be regarded as an array or table of numbers arranged in m rows and n columns. The same table of numbers could be arranged alternatively in n rows and m columns, with (for example) the sequence of numbers previously in the first column now written out as the first row. This new matrix of order $n \times m$ is called the transpose of the original matrix.

Definition. The *transpose* of a matrix $\mathbf{A} = [a_{ij}]$ of order $m \times n$ is a matrix obtained from **A** by interchanging rows and columns, so that row i of **A** becomes column i of the transposed matrix which is of order $n \times m$ and is denoted $\mathbf{A}'$.

The typical element of $\mathbf{A}'$ (i.e. the element in the ith row and j^{th} column) may similarly be denoted a'_{ij}. Because of the transposition this is the same as the element in the ith *column* and j^{th} *row* of **A**, which is denoted a_{ji}. Thus $a'_{ij} = a_{ji}$ and if we want to describe $\mathbf{A}'$ by reference to its typical element we may write

$$\mathbf{A}' = [a'_{ij}] = [a_{ji}]$$

The use of primed notation is consistent with (though not the same as) the distinction we have made between (for example) **b**, a column vector, and **b′**, the same series of elements expressed as a row vector, since the transpose of a column vector is a row vector.

Examples:

(1) If $\mathbf{A} = \begin{bmatrix} 1 & 2 \\ 3 & 4 \end{bmatrix}$ then $\mathbf{A}' = \begin{bmatrix} 1 & 3 \\ 2 & 4 \end{bmatrix}$

(2) If $\mathbf{A} = \begin{bmatrix} 1 & 2 \\ -3 & 0 \\ 2 & 4 \end{bmatrix}$ then $\mathbf{A}' = \begin{bmatrix} 1 & -3 & 2 \\ 2 & 0 & 4 \end{bmatrix}$

The operation of transposition has a number of general properties. Clearly $(\mathbf{A}')' = \mathbf{A}$ and $\mathbf{I}' = \mathbf{I}$. Three important properties are now proved as theorems.

Theorem 3.1. If matrices **A** and **B** are of order $m \times n$ and if $\mathbf{C} = \mathbf{A} + \mathbf{B}$, then $\mathbf{C}' = \mathbf{A}' + \mathbf{B}'$.

Proof. Consider the general element of $\mathbf{C}'$, i.e. the element of the ith row and j^{th} column:

$$\begin{aligned} c'_{ij} &= c_{ji} && \text{(since } \mathbf{C}' \text{ is transpose of } \mathbf{C}) \\ &= a_{ji} + b_{ji} && \text{(since } \mathbf{C} = \mathbf{A} + \mathbf{B}) \\ &= a'_{ij} + b'_{ij} && \text{(since } \mathbf{A}', \mathbf{B}' \text{ are transposes of } \mathbf{A}, \mathbf{B}) \end{aligned}$$

Thus the $(i, j)^{\text{th}}$ element of $\mathbf{C}'$ is equal to the $(i, j)^{\text{th}}$ element of $\mathbf{A}' + \mathbf{B}'$, and this is true for all i, j.

Thus $$\mathbf{C}' = \mathbf{A}' + \mathbf{B}'$$

(This result can obviously be extended to the sum of more than two matrices, and also to show that if $\mathbf{C} = \mathbf{A} - \mathbf{B}$, then $\mathbf{C}' = \mathbf{A}' - \mathbf{B}'$.)

Theorem 3.2. If matrices **A** and **B** are of order $m \times p$ and $p \times n$ respectively, then $(\mathbf{AB})' = \mathbf{B}'\mathbf{A}'$.

Proof. Notice that **AB** is defined; and also note that $\mathbf{B}'$ is of order $n \times p$ and $\mathbf{A}'$ is of order $p \times m$, and hence $\mathbf{B}'\mathbf{A}'$ is defined. Let $\mathbf{C} = \mathbf{AB}$ and consider the $(i, j)^{\text{th}}$ element of **C**:

$$
\begin{aligned}
c'_{ij} &= c_{ji} \\
&= \sum_{k=1}^{p} a_{jk}b_{ki} \quad \text{(from the definition of matrix multiplication)} \\
&= \sum_{k=1}^{p} a'_{kj}b'_{ik} = \sum_{k=1}^{p} b'_{ik}a'_{kj}
\end{aligned}
$$

which is the $(i, j)^{\text{th}}$ element of $\mathbf{B}'\mathbf{A}'$ (again from the definition of matrix multiplication). Thus the typical element of $(\mathbf{AB})'$ is the same as that of $\mathbf{B}'\mathbf{A}'$ and hence

$$(\mathbf{AB})' = \mathbf{B}'\mathbf{A}'$$

Theorem 3.3. If matrices $\mathbf{A}_1, \mathbf{A}_2, \ldots, \mathbf{A}_n$ are such that the product matrix $\mathbf{A}_1 \mathbf{A}_2 \ldots \mathbf{A}_n$ is defined, then $(\mathbf{A}_1 \mathbf{A}_2 \ldots \mathbf{A}_n)' = \mathbf{A}_n' \mathbf{A}_{n-1}' \ldots \mathbf{A}_2' \mathbf{A}_1'$.

Proof. Clearly this includes the result of Theorem 3.2 as a special case. The proof is by the method of *induction*: we will suppose that the result is true for some value of n, and then show that it must be true for $n+1$; finally since it is true for $n=2$, it must be true for $n=3$ and so on. First note that since $\mathbf{A}_1 \mathbf{A}_2 \ldots \mathbf{A}_n$ is defined, then so is $\mathbf{A}_n' \ldots \mathbf{A}_2' \mathbf{A}_1'$ (by a generalization of the argument used in the proof of Theorem 3.2). Next suppose that the result is true for some positive integer value for n; in other words suppose that

$$(\mathbf{A}_1 \mathbf{A}_2 \ldots \mathbf{A}_n)' = \mathbf{A}_n' \mathbf{A}_{n-1}' \ldots \mathbf{A}_2' \mathbf{A}_1'$$

Now introduce another matrix $\mathbf{A}_{n+1}$ of appropriate order

$$
\begin{aligned}
(\mathbf{A}_1 \mathbf{A}_2 \ldots \mathbf{A}_{n+1})' &= (\{\mathbf{A}_1 \mathbf{A}_2 \ldots \mathbf{A}_n\} \{\mathbf{A}_{n+1}\})' \\
&= \mathbf{A}'_{n+1} (\mathbf{A}_1 \mathbf{A}_2 \ldots \mathbf{A}_n)' \quad \text{(by Theorem 3.2)} \\
&= \mathbf{A}'_{n+1} \mathbf{A}'_n \ldots \mathbf{A}'_2 \mathbf{A}'_1
\end{aligned}
$$

Thus if the result holds true for n, it is true for $n+1$. But the result holds for $n=2$ (by Theorem 3.2) and hence it is true for $n=3$. Since it is true for $n=3$, it must be true for $n=4$, and so on. Thus the theorem is proved for all integer values of $n > 1$.

With this operation of transposition in mind, it is now convenient to introduce two more special types of matrix.

Definition. A square matrix $\mathbf{A}$ is said to be *symmetric* if $\mathbf{A} = \mathbf{A}'$.

Clearly, a non-square matrix cannot have this property since $\mathbf{A}$ and $\mathbf{A}'$ would be of different orders. Another way of expressing the condition $\mathbf{A} = \mathbf{A}'$ is to require $a_{ij} = a_{ji}$. Elements a_{ii} on the principal diagonal relate only to themselves, but each element a_{ij} on one side of the diagonal has the same value as its corresponding element a_{ji} on the other side.

Definition. A square matrix $\mathbf{A}$ is said to be *skew-symmetric* if $\mathbf{A} = -\mathbf{A}'$.

Here, the required condition (when expressed in terms of elements) is $a_{ij} = -a_{ji}$. For elements on the principal diagonal, we require $a_{ii} = -a_{ii}$. The only value which meets this requirement is $a_{ii} = 0$. Thus all elements on the principal diagonal of a skew-symmetric matrix are zero. An element on one side of the principal diagonal must be of the opposite sign but of the same magnitude as the corresponding element on the other side.

Examples:

(3) $\mathbf{A} = \begin{bmatrix} 2 & 3 & 7 \\ 3 & 1 & -5 \\ 7 & -5 & -4 \end{bmatrix}$ is symmetric and $\mathbf{B} = \begin{bmatrix} 0 & 1 & 2 \\ -1 & 0 & -3 \\ -2 & 3 & 0 \end{bmatrix}$ is skew-symmetric.

(4) Consider a *network* which is defined as a set of points or nodes where we suppose that each node is connected by direct links to each of one or more other nodes so that it is possible to travel from any node to any other node in the network. (In terms of a road network, think of the nodes as junctions, and the links as roads.) If the lengths of all the links are known, the network can be described very conveniently by a distance matrix $\mathbf{D}$ which is square, and which has its elements $d_{ij} \geqq 0$ defined as follows:

$d_{ii} = 0$

for $i \neq j$, $d_{ij} =$ length of link, if there exists a direct link from node i to node j

$d_{ij} = +\infty$, if there is no direct link from node i to node j (i.e. if it is possible to travel from i to j only by passing through another node.)

If the network is symmetric, i.e. if each link permits travel in both directions and the distance between the pair of nodes is the same in either direction, then (for $i \neq j$) $d_{ij} = d_{ji}$ where a link exists; where it does not, $d_{ij} = +\infty = d_{ji}$. Thus the matrix **D** is symmetric. If the network is not symmetric (e.g. if there are one-way streets, in the case of a road network) then **D** is not symmetric; note that it is not skew-symmetric either.

This distance matrix representation of the network is very useful for many types of network calculation, e.g. to find the shortest route from one node to another through the network.

3.15 *Exercises*

1. Find the product **AB** (a) by direct multiplication, and (b) by using the rule for the multiplication of partitioned matrices, where **A** and **B** are the following matrices, partitioned as shown:

$$\mathbf{A} = \left[\begin{array}{cc|c} 2 & 4 & -3 \\ 7 & 8 & 1 \\ \hline 6 & 0 & 1 \end{array}\right] \qquad \mathbf{B} = \left[\begin{array}{c|cc} 6 & 0 & 1 \\ 2 & 3 & -1 \\ \hline -4 & 5 & 0 \end{array}\right]$$

*2. For the matrices **A** and **B** shown below, find the product **AB** by partitioning the matrices in a way consistent with the following scheme, and then using the rule for the multiplication of partitioned matrices:

$$\mathbf{A} \text{ partitioned in the manner } \begin{bmatrix} \mathbf{A}_{11} & \mathbf{A}_{12} \\ \mathbf{A}_{21} & \mathbf{A}_{22} \end{bmatrix}$$

$$\mathbf{B} \text{ partitioned in the manner } \begin{bmatrix} \mathbf{B}_{11} \\ \mathbf{B}_{21} \end{bmatrix}$$

$$\mathbf{A} = \begin{bmatrix} 3 & 4 & 0 & 0 \\ 2 & 7 & 0 & 0 \\ -2 & 0 & 1 & 0 \\ 0 & -2 & 0 & 1 \end{bmatrix} \qquad \mathbf{B} = \begin{bmatrix} 4 & 5 \\ 1 & 0 \\ -2 & 1 \\ 2 & 0 \end{bmatrix}$$

3. Verify that $(\mathbf{AB})' = \mathbf{B}'\mathbf{A}'$ for the case where

$$\mathbf{A} = \begin{bmatrix} 2 & 7 & 3 \\ -1 & 0 & 4 \end{bmatrix} \quad \text{and} \quad \mathbf{B} = \begin{bmatrix} -2 & 3 \\ 4 & 5 \\ 0 & 2 \end{bmatrix}$$

*4. If **A** is any square matrix, prove that $\mathbf{A}+\mathbf{A}'$ is a symmetric matrix, and that $\mathbf{A}-\mathbf{A}'$ is skew-symmetric.

5. If **A** is any matrix, prove that **AA′** is defined and is a symmetric matrix.

*6. For the input–output model with n industries (as described in section 1.3), write out the system of equations in matrix and vector notation; then use an identity matrix in order to obtain a neat expression in which all the terms involving the variables $x_1, x_2, \ldots, x_n$ appear on one side of the equations.

*7. Let us return to the baker first considered in section 1.8. Since our last acquaintance, his business has prospered. He is making three products at each of two bakeries, and each product has three ingredients. The baker is planning the deliveries of these ingredients for the following day's production. Let a_{ij} denote the weight of the ith ingredient per dozen of the jth product, and let x_{jk} denote the number of dozens to be produced of the jth product at the kth bakery. These two sets of figures are given in the following tables:

	Weight of ingredients per dozen		
(a_{ij})	Loaves	Scones	Cakes
Flour (lb.)	4·8	0·5	0·3
Sugar (oz.)	0·5	0·5	1·0
Butter (oz.)	0·3	0·4	1·25

	No. of dozens to be produced at each bakery	
(x_{jk})	Old	New
Loaves	30	100
Scones	20	30
Cakes	10	10

By setting out the calculations in matrix terms, calculate the weights of each ingredient to be delivered to each bakery.

8. Write out in matrix and vector notation the restrictions in the example given in section 1.8.

9. Suppose that figures are available showing the value of a country's imports in a given period for each of 30 commodity groups which together comprise total import trade. Also available for each commodity group are estimates showing the proportion of imports of the commodity group going to each of 15 inland regions into which the country is divided. Devise appropriate notation and hence describe in matrix and vector terms how to calculate the total value of imports going to each inland region.
10. What interpretation can be placed on the property of matrix symmetry in the context of example (6) in section 3.4?
11. Show that any square matrix **A** can be written

$$\mathbf{A} = \frac{\mathbf{A}+\mathbf{A}'}{2} - \frac{\mathbf{A}-\mathbf{A}'}{2}$$

Consider this in the light of exercise 4 above.

*12. If the matrix **A** is skew-symmetric show that **B′AB** is also skew-symmetric, where **B** is any matrix for which the matrix product is defined.

13. Suppose that each of six people living in a village pass on village gossip to at least some of the others. Let these communication channels be represented by a matrix **A** of zero-one entries where

$a_{ij} = 1$ if person i passes on gossip to person j
$a_{ij} = 0$ if person i does not pass on gossip to person j.

Suppose that the matrix need not be symmetric. (What is the interpretation of this?)

If for these six people, the matrix is as shown below, which of the six people will eventually learn of a particular juicy scandal known initially only to person 1? (Assume that the scandal does not become known outside the group of six people.) What general comments can be made about the interpretation of this particular matrix?

$$\begin{bmatrix} 1 & 1 & 1 & 0 & 0 & 0 \\ 0 & 1 & 0 & 0 & 0 & 1 \\ 1 & 0 & 1 & 0 & 0 & 0 \\ 0 & 0 & 0 & 1 & 1 & 0 \\ 0 & 0 & 0 & 1 & 1 & 0 \\ 0 & 1 & 0 & 0 & 0 & 1 \end{bmatrix}$$

*14. Given that

$$\mathbf{A} = \begin{bmatrix} 8 & \alpha \\ 1 & \beta \end{bmatrix} \qquad \mathbf{B} = \begin{bmatrix} 1 & 0 \\ \beta & 1 \end{bmatrix}$$

$$\mathbf{C} = \begin{bmatrix} 3 & \alpha \\ 1 & \beta \end{bmatrix} \qquad \mathbf{D} = \begin{bmatrix} 2 & 0 \\ 0 & 2 \end{bmatrix}$$

for what values (if any) of α and β is it true that $\mathbf{A}+\mathbf{B} \geqq \mathbf{CD}$?

15. If **A**, **B**, **C** and **D** are as in the previous question, and if

$$\mathbf{E} = \begin{bmatrix} 0 & 0 \\ 0 & 0 \end{bmatrix} \quad \text{and } \mathbf{F} = \begin{bmatrix} 1 & \alpha \\ \beta & 1 \end{bmatrix}$$

for what values (if any) of α and β is it true that

$$\mathbf{ABCDEF} = \begin{bmatrix} 24 & 37 \\ 10 & 2 \end{bmatrix}?$$

CHAPTER 4

Elementary operations and the rank of a matrix

4.1 *Introduction*

This chapter deals with some further properties of matrices and with certain further operations which can be carried out on matrices. Its main function is to prepare the way for Chapter 5 (which considers whether a matrix operation can be defined which is analogous to the division of scalars) and for Chapter 6 (which discusses the very important subject of the solution of a set of linear equations). The first concept to be introduced in this chapter is that of *elementary operations*. In order to show one reason for an interest in such operations, consider for a moment the problem of solving a set of linear equations. To be specific, consider the following example of two equations in two unknowns:

$$\begin{aligned} 2x_1+5x_2 &= -11 \\ x_1-\ x_2 &= \quad 5 \end{aligned} \tag{4–1}$$

In order to solve this set, an elementary approach is sufficient: multiply the second equation by 2 to give

$$2x_1-2x_2=10$$

and then subtract this equation from the first equation of the initial set to give

$$7x_2=-21$$

Thus the solution is $\quad x_2=-\ 3$

and $\quad x_1=\quad 2$

The first two steps in this method of solution involved, respectively, multiplying an equation by a scalar, and subtracting one equation from another. Now consider the matrix of order 2×2

$$\begin{bmatrix} 2 & 5 \\ 1 & -1 \end{bmatrix}$$

which may be formed from the coefficients on the left-hand side of the equations (4–1). Operations rather like these first two steps of the solution can be carried out on the rows of this matrix: first multiply the second row by 2 to give

$$\begin{bmatrix} 2 & 5 \\ 2 & -2 \end{bmatrix}$$

and then subtract the second row from the first:

$$\begin{bmatrix} 0 & 7 \\ 2 & -2 \end{bmatrix}$$

These two steps are examples of *elementary row operations*, which are given a formal definition in the next section. The way in which such operations on the rows of the coefficient matrix can help in formal methods for the solution of sets of equations will be considered later, in Chapter 6.

4.2 *Elementary operations*

Definition. There are three types of *elementary row operations* which may be carried out on the rows of a matrix. These are

(i) the interchange of two rows
(ii) the multiplication of each element in a row by any scalar $\lambda \neq 0$
(iii) the addition, to each element of the jth row, of λ times the corresponding element of the ith row

For a given matrix **A**, it turns out that each of these types of operation can be carried out on the rows of **A** by premultiplying **A** by an appropriate matrix. This important result will first be illustrated by some examples, and then a more general argument will be given.

Examples:

(1) Suppose
$$\mathbf{A} = \begin{bmatrix} 2 & 7 \\ 1 & 2 \\ 3 & 0 \end{bmatrix}$$

If we want to interchange rows 1 and 3, it may be done by premultiplying **A** by a matrix denoted

$$\mathbf{E} = \begin{bmatrix} 0 & 0 & 1 \\ 0 & 1 & 0 \\ 1 & 0 & 0 \end{bmatrix}$$

since
$$\mathbf{EA} = \begin{bmatrix} 3 & 0 \\ 1 & 2 \\ 2 & 7 \end{bmatrix}$$

In other words, premultiplication by this matrix **E** is equivalent to interchanging the first and third rows of **A**. Notice that **E** may be obtained by carrying out the *same* elementary row operation on $\mathbf{I}_3$, the identity matrix of order 3.

(2) For the same matrix **A**, the operation of multiplying the second row by the scalar -7 can be done by premultiplying **A** by the matrix

$$\mathbf{E} = \begin{bmatrix} 1 & 0 & 0 \\ 0 & -7 & 0 \\ 0 & 0 & 1 \end{bmatrix}$$

Check the arithmetic for yourself. Notice again that the appropriate premultiplying matrix is obtained by carrying out the elementary row operation in question on the identity matrix of order 3.

(3) For the same matrix **A**, the operation of adding twice the first row to the second row to give a new second row may be done by premultiplication by

$$\mathbf{E} = \begin{bmatrix} 1 & 0 & 0 \\ 2 & 1 & 0 \\ 0 & 0 & 1 \end{bmatrix}$$

Check the arithmetic for yourself. Again **E** is obtained by carrying out this elementary row operation on $\mathbf{I}_3$.

These examples suggest some general properties of the premultiplying matrices, which can now be explored more formally. After carrying out an elementary row operation, we want to be left with a matrix which is of the same order as **A**; in other words **EA** must have the same order as **A**. If **A** is of order $m \times n$, then **E** must be of order $m \times m$. (In the examples above $m = 3$ and $n = 2$.) Inspection of the mechanics of the multiplications used in the examples suggests that it is generally true that any elementary row operation can be executed by premultiplication by a matrix formed by carrying out the specified row operation on an identity matrix of appropriate order. For each type of elementary row operation, this may be proved by arguments based closely on the definition of matrix multiplication. To illustrate such proofs, consider here only the second type of row operation. Suppose that the ith row is to be multiplied by $\lambda \neq 0$. If this operation is applied to $\mathbf{I}_m$, the premultiplying matrix is

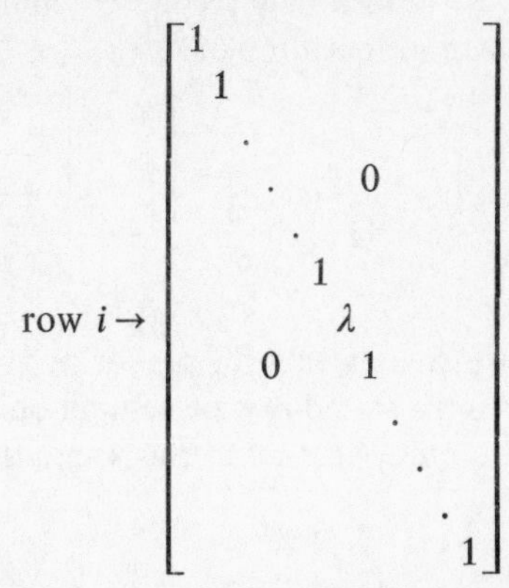

(Note that the zero elements off the principal diagonal are shown symbolically rather than in full detail.) For any row k other than row i, the multiplication of row k of this matrix into the jth column of $\mathbf{A}$ means that each element in the jth column is multiplied by zero except for the element which is in the kth row of the jth column, which is multiplied by unity; in other words all rows of $\mathbf{A}$ are left unaltered except for the ith row. In the case of the ith row, the multiplying element is λ and this yields the desired elementary row operation. In tracing the steps of this proof, consider example (2) above to see how it applies to a concrete instance.

Similar (but more complicated) proofs may be set out for the other two types of elementary row operation.

Elementary column operations may be defined similarly. They are equivalent to *postmultiplication* by matrices (of order $n \times n$, if $\mathbf{A}$ is of order $m \times n$) which are obtainable by carrying out the specified elementary column operations on the identity matrix of order n. All properties and proofs are precisely analogous to those of row operations.

Examples:

(4) If $\mathbf{A}$ is defined as in example (1), interchange of the first and second columns may be achieved by postmultiplication of $\mathbf{A}$:

$$\begin{bmatrix} 2 & 7 \\ 1 & 2 \\ 3 & 0 \end{bmatrix} \begin{bmatrix} 0 & 1 \\ 1 & 0 \end{bmatrix} = \begin{bmatrix} 7 & 2 \\ 2 & 1 \\ 0 & 3 \end{bmatrix}$$

(5) With the same matrix **A**, the addition of the first column to the second column to yield a new second column may be done by the following postmultiplication of **A**:

$$\begin{bmatrix} 2 & 7 \\ 1 & 2 \\ 3 & 0 \end{bmatrix} \begin{bmatrix} 1 & 1 \\ 0 & 1 \end{bmatrix} = \begin{bmatrix} 2 & 9 \\ 1 & 3 \\ 3 & 3 \end{bmatrix}$$

4.3 *Exercises*

In each of the following cases, find the pre- or postmultiplying matrix which corresponds to the stated row or column operations, and then carry out the matrix multiplication to check that the desired effect is achieved.

1. For the matrix $\mathbf{A} = \begin{bmatrix} 1 & 0 & 2 \\ 1 & 2 & 3 \\ 2 & 4 & -1 \end{bmatrix}$

   add three times the first row to the third row.
2. For the same matrix **A**, interchange the second and third columns.

*3. For the same matrix **A**, subtract the third row from the first row; in the new matrix subtract twice the third row from the second row. (First find the two separate premultiplying matrices which represent the two elementary row operations. Then multiply these two matrices together to obtain a single premultiplying matrix, and check that when this latter matrix is multiplied into **A**, the appropriate result is obtained.)

4. For the matrix $\mathbf{A} = \begin{bmatrix} 1 & 3 \\ 2 & 6 \\ -1 & -3 \end{bmatrix}$

   subtract six times the first column from twice the second column.

4.4 *Echelon matrices*

The third exercise in section 4.3 shows that we can find a single premultiplying matrix which corresponds to a sequence of *two* elementary row operations. However, it is often convenient to think of a sequence of elementary row operations as being represented by a similar sequence of premultiplying matrices, each of which corresponds to *one* elementary row operation and which hence may be called an *elementary matrix.*

Examples:

(1) Given the matrix $\mathbf{A} = \begin{bmatrix} 1 & 2 & 3 \\ 2 & 4 & 0 \\ 1 & 2 & 1 \end{bmatrix}$

consider the following sequence of elementary row operations:

(i) subtract twice row 1 from row 2
(ii) then add the new row 2 to row 3

By applying these operations to identity matrices of order 3, we see that the elementary matrices to represent the two operations are (respectively)

$$\mathbf{E}_1 = \begin{bmatrix} 1 & 0 & 0 \\ -2 & 1 & 0 \\ 0 & 0 & 1 \end{bmatrix} \quad \text{and} \quad \mathbf{E}_2 = \begin{bmatrix} 1 & 0 & 0 \\ 0 & 1 & 0 \\ 0 & 1 & 1 \end{bmatrix}$$

and thus the new matrix obtained by carrying out this sequence of operations on **A** is

$$\mathbf{E}_2\mathbf{E}_1\mathbf{A} = \mathbf{E}_2 \begin{bmatrix} 1 & 0 & 0 \\ -2 & 1 & 0 \\ 0 & 0 & 1 \end{bmatrix} \begin{bmatrix} 1 & 2 & 3 \\ 2 & 4 & 0 \\ 1 & 2 & 1 \end{bmatrix}$$

$$= \begin{bmatrix} 1 & 0 & 0 \\ 0 & 1 & 0 \\ 0 & 1 & 1 \end{bmatrix} \begin{bmatrix} 1 & 2 & 3 \\ 0 & 0 & -6 \\ 1 & 2 & 1 \end{bmatrix}$$

$$= \begin{bmatrix} 1 & 2 & 3 \\ 0 & 0 & -6 \\ 1 & 2 & -5 \end{bmatrix}$$

Note that in the product of the matrices, $\mathbf{E}_2$ is written down before $\mathbf{E}_1$ in order to represent correctly the sequence of operations.

(2) Given the matrix $\mathbf{A} = \begin{bmatrix} 1 & 2 & 3 \\ 1 & 2 & 4 \\ 0 & 0 & 1 \end{bmatrix}$

consider the following sequence of elementary row operations:

(i) subtract row 1 from row 2
(ii) subtract the new row 2 from row 3

This time $\mathbf{E}_1 = \begin{bmatrix} 1 & 0 & 0 \\ -1 & 1 & 0 \\ 0 & 0 & 1 \end{bmatrix}$ and $\mathbf{E}_2 = \begin{bmatrix} 1 & 0 & 0 \\ 0 & 1 & 0 \\ 0 & -1 & 1 \end{bmatrix}$

and so $$\mathbf{E}_2\mathbf{E}_1\mathbf{A} = \begin{bmatrix} 1 & 2 & 3 \\ 0 & 0 & 1 \\ 0 & 0 & 0 \end{bmatrix}$$

This last example illustrates the concept of an echelon matrix, whose definition is given next; we shall then see how this concept is related to the idea of a sequence of elementary row operations.

Definition. An *echelon matrix* is any matrix which has the following structure:

(i) each of the first k rows (where $k \geqq 0$) has one or more non-zero elements
(ii) for each such row, the first non-zero element (reading from left to right) is unity
(iii) the arrangement of these rows is such that the first non-zero element in a row is always to the right of the first non-zero element for each row higher up in the matrix
(iv) after the first k rows, the elements of the remaining rows (if any) are all zero.

The matrix $\mathbf{E}_2\mathbf{E}_1\mathbf{A}$ given at the end of example (2) above is clearly an example of an echelon matrix. Further examples are

$$\begin{bmatrix} 0 & 0 & 1 & -3 & 4 \\ 0 & 0 & 0 & 1 & 2 \\ 0 & 0 & 0 & 0 & 0 \end{bmatrix} \quad \text{and} \quad \begin{bmatrix} 1 & 2 & 0 & 4 \\ 0 & 1 & -7 & 3 \\ 0 & 0 & 1 & 2 \end{bmatrix}$$

This concept can now be used in further discussion of sequences of elementary row operations. We will show that it is always possible to find a sequence of elementary row operations which will transform a given matrix into an echelon matrix. (Note that we refer to 'an' echelon matrix; as will be seen below, it is not generally unique.) First consider an example in which the given matrix is as shown in Tableau 0 of Table 4.1. Bearing in mind the type of structure which characterizes an echelon matrix, it seems profitable to interchange

TABLE 4.1

Tableau 0			Tableau 1			Tableau 2			Tableau 3		
0	1	1	1	2	−1	1	2	−1	1	2	−1
1	2	−1	0	1	1	0	1	1	0	1	1
1	4	2	1	4	2	0	2	3	0	0	1

rows 1 and 2 as the first elementary row operation, leading to the matrix in Tableau 1. This gives a unit entry at the beginning of row 1. There is already a zero entry at the beginning of row 2, and if row 1 is now subtracted from row 3, we get a zero entry in row 3, as shown in the third matrix. For the third elementary row operation, subtract twice (the new) row 2 from row 3 to give the fourth matrix, which is now in echelon form. Thus for this particular example we have been able to find a sequence of elementary row operations which achieves the desired transformation. The general result is now stated formally:

Theorem 4.1. For any given matrix **A**, it is possible to find a sequence of elementary row operations which transforms **A** into an echelon matrix.

Proof. Here we merely outline the proof, which is of a type known as a 'constructive proof' which here means that the proof shows constructively how to achieve the transformation in the general case. Most of the ideas used in the general approach have been illustrated in the previous example. This general approach is to consider the first column in **A** which has at least one non-zero element; suppose this is column j. If, for this column, there is not a non-zero element in the first row, interchange that row with any later row which does have such an element (in the jth column). Suppose that the non-zero element (now) in the first row is λ; then divide that row by λ to give a unit element (as the first non-zero element in the first row). Then subtract appropriate multiples of this row from all later rows having non-zero elements (in this column) in order to obtain zero elements in the rest of the column. Then go on to consider the second row and the $(j+1)^{\text{th}}$

column (or more generally the next column having at least one non-zero element in any row from the second downwards); similar operations yield a unit entry in the second row and zero entries below. Continue in this way until the final column is reached; the matrix is then in echelon form.

Two points should be noted about this procedure. First, the echelon matrix is in *no* sense 'equal' to the given matrix. Instead, it is merely the case that the echelon matrix can be derived from the given matrix by applying a sequence of elementary row operations. Secondly, the procedure described in the proof of Theorem 4.1 is not unique and hence (in general) there is *not* a unique echelon matrix corresponding to a given matrix. This follows because in the procedure we speak of interchanging a row with a row lower down the matrix, and there may be a choice as to which lower row is used. (Again, we are entitled to interchange rows even when it is not essential to do so; *any* sequence of elementary row operations is valid provided we eventually reach an echelon matrix.) To illustrate the point, consider again the matrix used in Table 4.1. In that table, the first operation was to interchange rows 1 and 2. Suppose alternatively that we begin by interchanging rows 1 and 3, leading to the matrix in Tableau 0 of Table 4.2, and then elementary operations are carried out as in the remainder of that table. This yields an echelon matrix which is quite different from that of Table 4.1. Of course, in the general case, we could make the procedure unique by saying that rows should be interchanged only when necessary, and that the first suitable row lower down the matrix should be used in the interchange. Indeed, when a program is designed to enable an electronic computer to carry out the calculation, some such rule is employed. In hand

TABLE 4.2

Tableau 0			Tableau 1			Tableau 2			Tableau 3		
1	4	2	1	4	2	1	4	2	1	4	2
1	2	-1	0	-2	-3	0	1	$\frac{3}{2}$	0	1	$\frac{3}{2}$
0	1	1	0	1	1	0	1	1	0	0	1

calculation for small matrices, however, it is convenient to vary the procedure on occasion, to take advantage of any special structure of the given matrix, and hence to try to minimize the amount of computation.

4.5 *Exercises*

For each of the following matrices, carry out elementary row operations in order to find an echelon matrix.

*1. $\begin{bmatrix} 2 & 3 & 6 & 5 \\ 1 & 0 & 5 & 1 \\ 2 & 1 & 3 & 4 \end{bmatrix}$

2. $\begin{bmatrix} 1 & 3 & 2 \\ 2 & 6 & 4 \\ 1 & 0 & 1 \end{bmatrix}$

3. $\begin{bmatrix} 4 & 2 & 3 & 1 \\ 1 & 0 & 1 & 0 \\ 2 & 3 & 1 & 1 \\ 0 & 0 & 0 & 1 \end{bmatrix}$

4.6 *Elementary operations and linear dependence*

It is interesting to consider these elementary operations and the derivation of an echelon matrix in the light of the discussion (in Chapter 2) of linear combinations and linear dependence of vectors. To this end, let us now regard each row of the given matrix as a row vector. In transforming the given matrix into another matrix by a sequence of elementary row operations, each row in a matrix is expressed as a linear combination of one or more rows in the preceding matrix in the sequence; this is simply another way of looking at the definitions of the elementary row operations. As an example, take the derivation of an echelon matrix in Table 4.3 from the given matrix shown in Tableau 0 of that table. In considering the first elementary row operation, denote the ith row of the given matrix as $\mathbf{r}_i$ and that of the next matrix as $\mathbf{s}_i$. Then

$$\begin{aligned} \mathbf{s}_1 &= \mathbf{r}_1 \\ \mathbf{s}_2 &= \mathbf{r}_2 - 2\mathbf{r}_1 \\ \mathbf{s}_3 &= \mathbf{r}_3 - \mathbf{r}_1 \end{aligned}$$

and, as a convenient shorthand, these relationships may be noted at the side of Tableau 1 as a record of the operation carried out. Clearly these relationships are examples of what were defined (in section 2.5) as linear combinations, here of the row vectors $\mathbf{r}_1$, $\mathbf{r}_2$ and $\mathbf{r}_3$. This type of relationship holds every time an elementary row operation is carried out, and hence each row of the final matrix can be expressed as a linear combination of the rows of the initial matrix. For example, in Table 4.3, let the rows of the third matrix be denoted $\mathbf{t}_i$ and those of the echelon matrix $\mathbf{u}_i$. Then the remaining elementary

TABLE 4.3

Tableau 0				Tableau 1			
1	2	3		1	2	3	$\mathbf{r}_1$
2	4	0		0	0	−6	$\mathbf{r}_2-2\mathbf{r}_1$
1	2	1		0	0	−2	$\mathbf{r}_3-\mathbf{r}_1$
Tableau 2				**Tableau 3**			
1	2	3	$\mathbf{s}_1$	1	2	3	$\mathbf{t}_1$
0	0	1	$\mathbf{s}_2 \div -6$	0	0	1	$\mathbf{t}_2$
0	0	−2	$\mathbf{s}_3$	0	0	0	$\mathbf{t}_3+2\mathbf{t}_2$

row operations are as noted in the table. In particular, the third row of the echelon matrix is

$$\mathbf{u}_3 = \mathbf{t}_3+2\mathbf{t}_2$$

But $\mathbf{t}_3 = \mathbf{s}_3$ and $\mathbf{t}_2 = -\frac{1}{6}\mathbf{s}_2$ and hence

$$\mathbf{u}_3 = \mathbf{s}_3-\tfrac{1}{3}\mathbf{s}_2$$

From the relationships established above, this gives

$$\begin{aligned}\mathbf{u}_3 &= (\mathbf{r}_3-\mathbf{r}_1)-\tfrac{1}{3}(\mathbf{r}_2-2\mathbf{r}_1)\\ &= \mathbf{r}_3-\tfrac{1}{3}(\mathbf{r}_2+\mathbf{r}_1)\end{aligned}$$

In other words, the third row of the final (echelon) matrix is thus expressed as a linear combination of the rows of the given matrix.

These remarks about linear combinations apply to all elementary row operations; similar remarks apply to elementary column operations and linear combinations of column vectors. A further remark is of special interest in the context of echelon matrices: if an echelon matrix derived from a given matrix contains one or more rows of zero entries, then the rows of the given matrix must be linearly dependent. In Table 4.3, for example, since $\mathbf{u}_3$ is a row of zero entries, then from the above relationship

$$\mathbf{r}_3 - \tfrac{1}{3}(\mathbf{r}_2 + \mathbf{r}_1) = \mathbf{0}$$

i.e. $$[1 \quad 2 \quad 1] - \tfrac{1}{3}[2 \quad 4 \quad 0] - \tfrac{1}{3}[1 \quad 2 \quad 3] = [0 \quad 0 \quad 0]$$

Check for yourself that the arithmetic is correct.

4.7 *Exercises*

For each of the following matrices, carry out elementary row operations in order to find an echelon matrix, setting out your working in the style of Table 4.3. In each case, you should find that the rows of the given matrix are linearly dependent; write down a vector equation which expresses this linear dependence (i.e. an equation in the form

$$\lambda_1\mathbf{r}_1 + \lambda_2\mathbf{r}_2 + \ldots + \lambda_n\mathbf{r}_n = \mathbf{0}$$

where the $\mathbf{r}_i$ are the rows of the given matrix and the λ_i are scalars you must calculate).

*1. $\begin{bmatrix} 1 & 2 \\ 3 & 4 \\ 0 & 2 \end{bmatrix}$

2. $\begin{bmatrix} 2 & 4 & 6 \\ 2 & 5 & 0 \\ 4 & 9 & 6 \end{bmatrix}$

4.8 *The rank of a matrix*

In section 4.6 a matrix was regarded as a set of row vectors; this in turn now leads to consideration of whether the rows of a given matrix are linearly dependent or independent. As will be seen in later chapters, this is a very important and useful distinction; accordingly we now give some definitions which introduce this subject.

Definition. The *rank* of a rectangular matrix is defined as the maximum number of linearly independent rows in the matrix.

Just as in section 2.5 where we also considered sets of vectors, the idea here is to consider all possible sets of row vectors which may be formed by selecting from the entire set of row vectors which make up the matrix. If the largest linearly independent set which can be found comprises k vectors, then we say that the rank of the matrix is k.

Examples:

(1) If the given matrix is $\mathbf{A} = \begin{bmatrix} 2 & 7 \\ 1 & 3 \end{bmatrix}$

then since there are only two vectors, we know immediately that $k \leqq 2$. Let us test first to see if the two rows are linearly independent. Suppose that there exist scalars λ_1 and λ_2 such that

$$\lambda_1[2 \quad 7] + \lambda_2[1 \quad 3] = [0 \quad 0]$$

Then $$2\lambda_1 + \lambda_2 = 0$$

and $$7\lambda_1 + 3\lambda_2 = 0$$

The only solution we can find is $\lambda_1 = \lambda_2 = 0$. Thus the row vectors are linearly independent (cf. section 2.5), and hence the rank of $\mathbf{A}$ is $k = 2$.

(2) Consider $$\mathbf{B} = \begin{bmatrix} 2 & 7 \\ 1 & 3 \\ 0 & 2 \end{bmatrix}$$

This time there are three rows; but each vector has 2 components, i.e. it can be considered as a vector from E^2. By applying Theorem 2.6 we know that three vectors from E^2 must be linearly dependent. Thus the rank cannot exceed 2. Now there are three possible ways of forming a set of two row vectors from the given set of three rows. And we already know something about one of these: if we drop the third row, we get the set of row vectors just considered in example (1), and we know this set to be linearly independent. Thus the largest set of linearly independent rows in $\mathbf{B}$ comprises two vectors; hence the rank of $\mathbf{B}$ is $k = 2$.

(3) Consider $$\mathbf{C} = \begin{bmatrix} 1 & 3 \\ 2 & 6 \\ 3 & 9 \end{bmatrix}$$

As in example (2), we know that three vectors from E^2 must be linearly dependent. Thus the rank cannot exceed 2. Consider in turn the three sets each of two vectors: (a) [1 3] and [2 6]; (b) [1 3] and [3 9]; (c) [2 6] and [3 9]. In each of the three cases, it is clear that one vector is a multiple of the other, and hence that there exist scalars λ_1 and λ_2 (not both equal to zero) such that $\lambda_1\mathbf{r}+\lambda_2\mathbf{s}=\mathbf{0}$ where **r** and **s** denote the two row vectors; in other words, each set of two vectors is linearly dependent. Thus the rank cannot exceed one. Consider now (for example) the first row. This is a linearly independent set (of one vector) since

$$\lambda_1[1 \quad 3]=[0 \quad 0]$$

is satisfied only for $\lambda_1=0$. Thus the rank of **C** is $k=1$. (And, in general, we see from this last part of the argument that any matrix with a non-zero element – i.e. any matrix other than a null matrix – has rank not less than one.)

In these examples, we have used very elementary ways of calculating the rank. As will be seen shortly, a more formal technique is available; and such a technique is essential when large matrices are to be examined.

Now in earlier discussion of matrices, we have witnessed a certain amount of symmetry as between rows and columns of a matrix. The definition of rank has been given in terms of rows. Suppose instead we considered the alternative definition in terms of columns. In order to explore this, first consider again the three examples. For the matrix **A**, suppose there exist scalars ϕ_1 and ϕ_2 such that

$$\phi_1\begin{bmatrix}2\\1\end{bmatrix}+\phi_2\begin{bmatrix}7\\3\end{bmatrix}=\begin{bmatrix}0\\0\end{bmatrix}$$

This gives the pair of equations

$$\begin{aligned}2\phi_1+7\phi_2&=0\\ \phi_1+3\phi_2&=0\end{aligned}$$

and the only solution is $\phi_1=\phi_2=0$. Thus the columns of **A** are linearly independent and the (column) rank of **A** is 2, which is the same as the rank (as defined originally in terms of rows). Similarly we find that the (column) rank of **B** is 2; and that that of **C** is 1, since

$$3\begin{bmatrix}1\\2\\3\end{bmatrix}+(-1)\begin{bmatrix}3\\6\\9\end{bmatrix}=\begin{bmatrix}0\\0\\0\end{bmatrix}$$

These examples suggest that the two definitions always leads to the same measure. In fact this result does hold generally. Although it is of considerable interest, we shall not have much occasion to use it; it is merely stated here formally, without proof:

Theorem 4.2. In a given rectangular matrix, the maximum number of linearly independent rows is equal to the maximum number of linearly independent columns.

It follows from this theorem that we may compute the rank of a matrix by finding the maximum number of linearly independent rows *or* the maximum number of linearly independent columns.

This definition of rank may be applied to all matrices, whether square or rectangular. When attention is confined to *square* matrices, a related definition is used to classify matrices according to rank:

Definition. A square matrix of order n is said to be *singular* if the rank is less than n, and to be *non-singular* if the rank equals n.

Note that these terms are applied only to square matrices. In the previous examples, $\mathbf{A}$ is non-singular. The matrix $\begin{bmatrix} 2 & 4 \\ 1 & 2 \end{bmatrix}$ is singular, since its rank is 1, as may readily be calculated.

Example:

(4) $\mathbf{I}_n$ has rank n, since the rows are linearly independent; hence it is a non-singular matrix.

4.9 *Product matrices and rank*

If two matrices $\mathbf{A}$ and $\mathbf{B}$ have ranks which are denoted $r(\mathbf{A})$ and $r(\mathbf{B})$, what is the rank of the product matrix $\mathbf{AB}$? In order to study this question, consider first the following examples, in each case checking for yourself the arithmetic of the matrix multiplication and of the derivation of the rank of $\mathbf{AB}$, denoted $r(\mathbf{AB})$:

Examples:

(1) $\mathbf{A} = \begin{bmatrix} 2 & 1 & 0 \\ 4 & 3 & 1 \end{bmatrix}, \quad \mathbf{B} = \begin{bmatrix} 1 & 1 \\ 2 & 1 \\ 3 & 2 \end{bmatrix}, \quad \mathbf{AB} = \begin{bmatrix} 4 & 3 \\ 13 & 9 \end{bmatrix}$

$r(\mathbf{A}) = 2 \qquad r(\mathbf{B}) = 2 \qquad r(\mathbf{AB}) = 2$

(2) $\mathbf{A} = \begin{bmatrix} 2 & 1 & 0 \\ 4 & 2 & 0 \end{bmatrix}$, $\mathbf{B}$ as before, $\mathbf{AB} = \begin{bmatrix} 4 & 3 \\ 8 & 6 \end{bmatrix}$

$r(\mathbf{A}) = 1 \qquad r(\mathbf{B}) = 2 \qquad r(\mathbf{AB}) = 1$

(3) $\mathbf{A} = \begin{bmatrix} 2 & 1 \\ 4 & 2 \end{bmatrix}$, $\mathbf{B} = \begin{bmatrix} 1 & -3 \\ -2 & 6 \end{bmatrix}$, $\mathbf{AB} = \begin{bmatrix} 0 & 0 \\ 0 & 0 \end{bmatrix}$

$r(\mathbf{A}) = 1 \qquad r(\mathbf{B}) = 1 \qquad r(\mathbf{AB}) = 0$

(4) $\mathbf{A} = \begin{bmatrix} 2 & 7 \\ 1 & 3 \\ 0 & 2 \end{bmatrix}$, $\mathbf{B} = \begin{bmatrix} 1 & 2 & 3 \\ 4 & 5 & 6 \end{bmatrix}$

$$\mathbf{AB} = \begin{bmatrix} 30 & 39 & 48 \\ 13 & 17 & 21 \\ 8 & 10 & 12 \end{bmatrix}$$

$r(\mathbf{A}) = 2 \qquad r(\mathbf{B}) = 2$

and $r(\mathbf{AB}) = 2$ since, denoting rows of $\mathbf{AB}$ by $\mathbf{r}_i$, we note

$$2\mathbf{r}_1 - 4\mathbf{r}_2 - \mathbf{r}_3 = \mathbf{0}$$

(showing that $r(\mathbf{AB}) < 3$), and also the first two rows (for example) are linearly independent.

(5) $\mathbf{A} = \begin{bmatrix} 1 & 3 \\ 2 & 1 \end{bmatrix}$, $\mathbf{B} = \begin{bmatrix} 1 & 2 & 0 \\ 2 & 3 & 1 \end{bmatrix}$

$r(\mathbf{A}) = 2 \qquad r(\mathbf{B}) = 2 \qquad r(\mathbf{AB}) = 2$

In each of these examples we note that $r(\mathbf{AB})$ is not greater than the smaller of the ranks of $\mathbf{A}$ and $\mathbf{B}$. Also we know in general (from the definition of rank and from Theorem 4.2) that the rank of a matrix cannot exceed the number of rows or the number of columns, whichever is smaller. From this last observation, in example (4), for instance, the rank of $\mathbf{AB}$ could be as great as three; but it turns out to be only two. The moral to be drawn from these examples is in fact quite general, and is the subject of the next theorem.

Theorem 4.3. If two matrices $\mathbf{A}$, $\mathbf{B}$ have ranks denoted $r(\mathbf{A})$ and $r(\mathbf{B})$ respectively, and are such that the product matrix $\mathbf{AB}$ is defined, then the rank of $\mathbf{AB}$, denoted $r(\mathbf{AB})$, satisfies the following inequality:

$$r(\mathbf{AB}) \leqq \min\,[r(\mathbf{A}), r(\mathbf{B})]$$

In other words, the rank of **AB** is not greater than the lesser of the ranks of **A** and **B**.

Proof. Suppose that **A** is of order $m \times p$ and **B** is of order $p \times n$. Then let **C** denote the product matrix **AB**; **C** must be of order $m \times n$. Now consider **B** as being composed of n column vectors $\mathbf{b}_j$ each of order p:

$$\mathbf{B} = [\mathbf{b}_1 \quad \mathbf{b}_2 \ldots . \mathbf{b}_n]$$

Similarly regard **A** as being composed of m *row* vectors $\mathbf{a}_i'$, each of order p, and also regard **C** as being composed of n column vectors $\mathbf{c}_j$ each of order m.

With this vector notation, we may write

$$\mathbf{C} = \mathbf{AB} = [\mathbf{c}_1 \quad \mathbf{c}_2 \ldots . \mathbf{c}_n] = \begin{bmatrix} \mathbf{a}_1'\mathbf{b}_1 & \mathbf{a}_1'\mathbf{b}_2 \ldots . \mathbf{a}_1'\mathbf{b}_n \\ \cdot & \\ \cdot & \\ \cdot & \\ \cdot & \\ \mathbf{a}_m'\mathbf{b}_1 & \mathbf{a}_m'\mathbf{b}_2 \ldots . \mathbf{a}_m'\mathbf{b}_n \end{bmatrix}$$

after using the usual rule for multiplying matrices.

It will first be shown that $r(\mathbf{C})$ cannot exceed $r(\mathbf{B})$. Observe that $r(\mathbf{B}) \leqq n$ since the rank of a matrix cannot exceed the number of columns in the matrix. Two cases must now be distinguished. First suppose that $r(\mathbf{B}) = n$. Since **C** has n columns, $r(\mathbf{C}) \leqq n$ and hence $r(\mathbf{C}) \leqq r(\mathbf{B})$; in other words, the result follows immediately in this case.

Secondly, suppose the alternative case i.e. $r(\mathbf{B}) < n$. Now consider *any* set of k columns from **C** where $n \geqq k > r(\mathbf{B})$. Without loss of generality, suppose that these are the *first* k vectors in **C**.

The *corresponding* set of k column vectors from **B** must be linearly dependent (since $r(\mathbf{B}) < k$), and hence there exist scalars λ_i (for $i = 1, 2, \ldots, k$) not all zero, such that

$$\lambda_1\mathbf{b}_1 + \lambda_2\mathbf{b}_2 + \ldots . + \lambda_k\mathbf{b}_k = \mathbf{0}$$

Now form a system of equations by premultiplying this equation in turn by each $\mathbf{a}_i'$ $(i = 1, 2, \ldots, m)$:

$$\lambda_1 \mathbf{a}_1'\mathbf{b}_1 + \lambda_2 \mathbf{a}_1'\mathbf{b}_2 + \ldots + \lambda_k \mathbf{a}_1'\mathbf{b}_k = \mathbf{0}$$
$$\vdots$$
$$\lambda_1 \mathbf{a}_m'\mathbf{b}_1 + \lambda_2 \mathbf{a}_m'\mathbf{b}_2 + \ldots + \lambda_k \mathbf{a}_m'\mathbf{b}_k = \mathbf{0}$$

This system of equations may be written as a single vector equation:

$$\lambda_1 \begin{bmatrix} \mathbf{a}_1'\mathbf{b}_1 \\ \cdot \\ \cdot \\ \cdot \\ \cdot \\ \mathbf{a}_m'\mathbf{b}_1 \end{bmatrix} + \lambda_2 \begin{bmatrix} \mathbf{a}_1'\mathbf{b}_2 \\ \cdot \\ \cdot \\ \cdot \\ \cdot \\ \mathbf{a}_m'\mathbf{b}_2 \end{bmatrix} + \ldots + \lambda_k \begin{bmatrix} \mathbf{a}_1'\mathbf{b}_k \\ \cdot \\ \cdot \\ \cdot \\ \cdot \\ \mathbf{a}_m'\mathbf{b}_k \end{bmatrix} = \mathbf{0}$$

But these k column vectors are the first k columns of **C**. And because at least one $\lambda_i \neq 0$, then these k vectors are linearly dependent. Thus we have proved that any k columns from **C** are linearly dependent, where k is any number greater than $r(\mathbf{B})$. In other words, the rank of **C** cannot exceed $r(\mathbf{B})$. Similarly, by regarding **C** as being composed of m rows, it may be shown that the rank of **C** cannot exceed $r(\mathbf{A})$. The proof follows precisely the same pattern as before, except that the linear dependence relation between t rows of **A** (where $t > r(\mathbf{A})$) is *postmultiplied* in turn by each of the column vectors $\mathbf{b}_j$ for $j = 1, 2, \ldots, n$.

When these two results are combined, we have

$$r(\mathbf{C}) = r(\mathbf{AB}) \leqq \min\,[r(\mathbf{A}), r(\mathbf{B})]$$

At this point, it is convenient to state a further theorem, even though we do not give a proof at this stage. This theorem can be regarded as a special case of the previous theorem, and one in which the equality sign holds. The theorem describes what happens when a matrix is premultiplied or postmultiplied by a *non-singular* matrix. Consider premultiplication. Let **B** denote a matrix of order $m \times n$, having rank k (where, of course, $k \leqq \min\,(m, n)$ since the rank cannot exceed the lesser of the number of rows and the number of columns). Suppose **B** is premultiplied by a non-singular matrix **A**; by implication, **A** must be square of order m, with rank m. Example (5) above illustrates what happens; in this example $m = 2$, $n = 3$ and $k = 2$. From Theorem 4.3, we know that $r(\mathbf{AB}) \leqq 2$. As the working of the example shows, $r(\mathbf{AB})$ in fact *equals* 2. In other words, $r(\mathbf{AB}) = r(\mathbf{B})$, and this type of result holds generally, as indicated in the following theorem.

Theorem 4.4. If a matrix of rank k is premultiplied (or postmultiplied) by a non-singular matrix, the rank of the product matrix is k.

A very simple proof of this theorem can be given once we have introduced some further concepts at the beginning of Chapter 5. (The theorem will not be used in the derivation of further theorems until after that stage.)

4.10 *Computing the rank of a matrix*

In section 4.8, very elementary methods were employed to calculate the ranks of some small matrices. For larger matrices a more systematic approach is required; a very efficient technique is developed in this section. The basic idea is to apply elementary row operations in order to derive an echelon matrix from the matrix whose rank is to be calculated; the rank of the echelon matrix can then be ascertained immediately, and it is the same as that of the given matrix. The theorems necessary to support this argument are given next.

Theorem 4.5. If **B** is a matrix obtained by applying an elementary row operation to a given matrix **A**, then $r(\mathbf{B}) = r(\mathbf{A})$.

Proof. Consider the first type of elementary row operation, the interchange of two rows. This leaves unaltered the set of row vectors making up the matrix **A**, and simply alters the order in which these vectors are listed in order to give **B**. Thus the *size* of the largest set of linearly independent row vectors in **A** is the same as the size of the largest such set in **B**.

Similar arguments apply to the second and third types of elementary row operation. In the second case, for example, one row vector in **A** is multiplied by a scalar to give the corresponding row in **B**; but this does not change the linear dependence and linear independence relationships; these are just the same in **B** as they are in **A**.

Thus in all three cases, $r(\mathbf{B}) = r(\mathbf{A})$.

Theorem 4.6. An echelon matrix obtained by applying elementary row operations to a given matrix **A** has the same rank as **A**.

Proof. Note that **A** need not be a square matrix. Let $\mathbf{E}_1$ denote the matrix representing the first of the sequence of p elementary row operations which transforms **A** into an echelon matrix. Then $\mathbf{E}_1\mathbf{A}$ has the same rank as **A**, by Theorem 4.5. If $\mathbf{E}_2$ represents the second elementary row operation in the sequence, $\mathbf{E}_2(\mathbf{E}_1\mathbf{A})$ has the same rank as $\mathbf{E}_1\mathbf{A}$ (by Theorem 4.5) and hence the same rank as **A**. Repeated further application of this argument shows that the eventual echelon matrix $\mathbf{E}_p \ldots . \mathbf{E}_3\mathbf{E}_2\mathbf{E}_1\mathbf{A}$ has the same rank as **A**.

Theorem 4.7. A matrix representing any sequence of elementary row operations is non-singular.

Proof. Consider again the three types of elementary row operation. It may readily be shown that each of the three types of matrix representing these operations is itself non-singular. For instance, consider the first type of operation, the interchange of two rows. If this is applied to rows i and j of a matrix of order n, the corresponding matrix representation of the operation is an identity matrix of order n, with rows i and j interchanged. Clearly the rows of this matrix are linearly independent. (Test this for yourself in the usual way, by trying to find scalars not all zero which make a linear combination of these rows equal to a null row vector.) In other words the matrix is of rank n, and is non-singular.

The proof may now be completed by using Theorem 4.5 in the same manner as in the proof of Theorem 4.6. If the sequence of p elementary row operations is denoted by a sequence of premultiplying matrices $\mathbf{E}_p \ldots . \mathbf{E}_2\mathbf{E}_1$, then since $\mathbf{E}_1$ is non-singular (and of order n), then from Theorem 4.5, $\mathbf{E}_2\mathbf{E}_1$ is also non-singular (and of the same order of course). Theorem 4.5 may be applied again to show that $\mathbf{E}_3(\mathbf{E}_2\mathbf{E}_1)$ is non-singular, and so on until the entire product matrix $\mathbf{E}_p \ldots . \mathbf{E}_2\mathbf{E}_1$ is seen to be non-singular.

Theorem 4.8. The rank of an echelon matrix is equal to the number of rows comprising non-zero entries.

Proof. Let **G** denote an echelon matrix having k rows with non-zero entries. First, the rank of **G** cannot be greater than k, since any set of row vectors comprising more than k rows will contain at least one row of zero entries (i.e. a null vector) and thus must be linearly dependent; for this result, consider again exercise 5 of section 2.6 or work it out from first principles. Thus the maximum number of linearly independent rows can be no greater than k.

Secondly, we will now show that the first k rows (which are the rows with non-zero entries) *are* linearly independent. (Test the following argument on an example of an echelon matrix.) Let these rows be denoted $\mathbf{g}_1, \mathbf{g}_2, \ldots, \mathbf{g}_k$, and let us now try to find scalars λ_i such that

$$\lambda_1\mathbf{g}_1 + \lambda_2\mathbf{g}_2 + \ldots + \lambda_k\mathbf{g}_k = \mathbf{0}$$

Consider the column containing the first non-zero element (a unit element) of $\mathbf{g}_1$; all the other rows in the set have zero entries in this column; thus the corresponding equation from the set of equations in the scalars is

$$\lambda_1(1) + \lambda_2(0) + \ldots + \lambda_k(0) = 0$$
$$\therefore \qquad \lambda_1 = 0$$

Similar arguments show that $\lambda_2 = \lambda_3 = \ldots = \lambda_{k-1} = 0$. The vector equation then becomes

$$\lambda_k\mathbf{g}_k = \mathbf{0}$$

where $\mathbf{g}_k$ has at least one non-zero entry. Thus the only solution is $\lambda_k = 0$. Thus the set of k row vectors is linearly independent.

From the first part of the proof, every set of $(k+1)$ rows is linearly dependent. Thus the maximum number of linearly independent rows is k, and hence the rank is k.

These theorems permit the rank to be calculated in a very efficient way: reduce the given matrix to an echelon matrix and then in the echelon count the number of rows which have non-zero entries. Although this is one of the most efficient procedures for computing the rank of a matrix, finding the rank of a large matrix is generally not a trivial task.

Examples:

(1) The rank of the matrix shown in Tableau 0 of Table 4.3 (in section 4.6) is 2, because the echelon matrix shown in Tableau 3 has 2 rows with non-zero entries.

(2) The rank of the matrix shown in Tableau 0 of Table 4.2 (in section 4.4) is 3, because the echelon matrix shown in Tableau 3 has non-zero entries in all 3 rows. Note that this matrix is non-singular since it is a square matrix whose rank is the same as its order. Note also that the echelon matrix has unit entries in all positions on the principal diagonal, and that this must be so whenever an echelon matrix is derived from a non-singular matrix (since the echelon has no rows comprising zero entries).

4.11 *Exercises*

1. By obtaining an echelon matrix in each case, find the rank of

$$\mathbf{A} = \begin{bmatrix} 2 & 3 \\ 1 & 2 \\ 4 & 6 \end{bmatrix} \quad \text{and of } \mathbf{B} = \begin{bmatrix} 1 & 2 & 3 \\ 4 & 5 & 6 \\ 7 & 8 & 9 \end{bmatrix}$$

*2. Find an echelon matrix derivable from **BA**, where **B** and **A** are as defined in exercise 1, and hence compare the rank of **BA** with the ranks of **B** and **A**. Consider the result in the light of Theorem 4.3.

*3. If

$$\mathbf{A} = \begin{bmatrix} 1 & 1 & 0 \\ 1 & 2 & 0 \\ 1 & 2 & 1 \end{bmatrix} \quad \text{and } \mathbf{B} = \begin{bmatrix} 1 & 2 & \alpha \\ 2 & \beta & 3 \\ 4 & 8 & 6 \end{bmatrix}$$

for what values of α and β (if any) is the rank of **AB** equal to one? (Hint: a cunning way of tackling this is to find the rank of **A** and then employ Theorem 4.4.)

4. Under what circumstances (if any) is a diagonal matrix singular?

*5. What is the rank of the matrix $\mathbf{C} = \mathbf{A}+\mathbf{B}$, if **A** and **B** are square matrices of order n, with typical elements $a_{ij} = i+j$ and $b_{ij} = i-j$?

CHAPTER 5

The inverse of a square matrix

5.1 *The concept of an inverse matrix*

In Chapter 3, matrix operations of addition, subtraction and multiplication were defined. In arithmetic and in scalar algebra, there is also an operation of division. Can a similar operation be defined for matrices? The present chapter deals at some length with this question. The answer is that such an operation can be defined, but the parallel with the division of scalars is by no means exact. In fact the differences are so great that it is better to think in terms of comparing the matrix operation with the scalar operation of multiplying by a reciprocal. In arithmetic, instead of dividing by 2 we can speak of multiplying by $(2)^{-1}$. More generally, given a scalar $\lambda \neq 0$, we can speak of multiplying by λ^{-1}; this operation has the property

$$\lambda^{-1}\lambda = \lambda\lambda^{-1} = 1$$

For matrix algebra, this prompts the question: for a given matrix **A** can a matrix **B** be found such that

$$\mathbf{BA} = \mathbf{AB} = \mathbf{I}$$

where **I** is an identity matrix of order n? For these matrix equations to hold, **B** must have n rows and **A** must have n columns, so that **BA** is of order $n \times n$. Similarly **A** must have n rows and **B** must have n columns if **AB** is to be of order $n \times n$. In other words if we want to confine our attention to cases where **BA** and **AB** are of the same order, then **A** must be a square matrix, in which case **B** is also square. This leads to the following definition of an inverse matrix (which bears a rough analogy to the reciprocal of scalar algebra):

Definition. Given a square matrix **A**, if there exists a square matrix, to be denoted $\mathbf{A}^{-1}$, which satisfies the relation

$$\mathbf{A}^{-1}\mathbf{A} = \mathbf{A}\mathbf{A}^{-1} = \mathbf{I}$$

then $\mathbf{A}^{-1}$ is called the *inverse matrix* (or simply the *inverse*) of **A**.

Notice that the definition has to be qualified by the clause 'if there exists a square matrix'. Nothing that has been said so far demonstrates that such a matrix will exist for a given (square) matrix **A**. Indeed we shall see later that in some circumstances no such matrix exists, i.e. that it is not possible to find a matrix $\mathbf{A}^{-1}$ which satisfies the definition.

Further note that the definition requires the *same* square matrix for both pre- and postmultiplication. According to this definition, it is not sufficient to find a matrix **B** such that $\mathbf{BA} = \mathbf{I}$ and a matrix **C** such that $\mathbf{AC} = \mathbf{I}$, unless $\mathbf{B} = \mathbf{C}$. In other words we want an inverse which commutes with **A**. (This aspect will be more thoroughly explored below in section 5.4.) Finally observe that in adopting this definition, we ignore the question of whether we can define some sort of inverse when the given matrix **A** is *not* square. In other words, this chapter deals only with square matrices.

Example:

Given a matrix $\mathbf{A} = \begin{bmatrix} 1 & 2 \\ 1 & 3 \end{bmatrix}$

then the matrix $\mathbf{B} = \begin{bmatrix} 3 & -2 \\ -1 & 1 \end{bmatrix}$

satisfies the relation $\mathbf{BA} = \mathbf{AB} = \mathbf{I}_2$, and hence **B** is the inverse of **A**. (Check this for yourself by computing the product matrices **BA** and **AB**.) Since **B** satisfies the relation, an inverse of **A** exists in this case. More generally, in any case where it is possible to find a matrix with the appropriate properties, then the existence of an inverse is thereby demonstrated for that case.

5.2 *An approach to calculating the inverse matrix*

In this section, let us suppose that, given a square matrix **A** of order n, there exists a matrix **B** which satisfies *only* the premultiplication requirement, namely that

$$\mathbf{BA} = \mathbf{I}_n$$

where **B** is, of course, also of order n. For the present, we do *not* suppose that **B** also satisfies the postmultiplication requirement.

Considering the above equation, the task of finding **B** can be thought of as that of finding a premultiplying matrix which transforms **A** into the identity matrix $\mathbf{I}_n$. This is reminiscent of the procedure discussed in the previous chapter for applying a sequence of elementary row operations (which can be represented by a premultiplying matrix) in order to transform the given matrix into an echelon matrix. *If* we can find a sequence of elementary row operations which transforms **A** into $\mathbf{I}_n$, then the premultiplying matrix which represents this sequence must be the matrix **B** we are looking for, given the assumption made in this section about its existence. Can such a sequence be found? Before proving in a later section that it can, the rest of this section will be devoted to the exploration of an example, in order to see whether and how the idea works in a simple case.

Suppose that the given square matrix is $\mathbf{A} = \begin{bmatrix} 1 & 1 \\ 2 & 3 \end{bmatrix}$ and, in the spirit of this section, suppose that for this matrix there exists a square matrix **B** such that $\mathbf{BA} = \mathbf{I}_2$. The first part of the procedure is to find a sequence of elementary row operations which transforms the given matrix to an echelon matrix; the discussion in the previous chapter has shown that this can always be achieved. In the present case it is simply done: subtract twice the first row from the second row, i.e. premultiply by

$$\mathbf{E}_1 = \begin{bmatrix} 1 & 0 \\ -2 & 1 \end{bmatrix}, \text{ and the result is } \mathbf{E}_1\mathbf{A} = \begin{bmatrix} 1 & 1 \\ 0 & 1 \end{bmatrix}$$

Note that this is the kind of echelon matrix which does not have any rows exclusively composed of zero entries, and hence there are unit entries all the way down the principal diagonal. So far, the elementary row operations have served to create zero entries everywhere below the principal diagonal. If the matrix is to be further transformed in order to reach an identity matrix, the remaining task is to find further elementary row operations in order to secure zero entries above the principal diagonal. Again, in the present case, this is very simply done: subtract the second row from the first, i.e. premultiply by

$$\mathbf{E}_2 = \begin{bmatrix} 1 & -1 \\ 0 & 1 \end{bmatrix}, \text{ and the result is } \mathbf{E}_2\mathbf{E}_1\mathbf{A} = \begin{bmatrix} 1 & 0 \\ 0 & 1 \end{bmatrix}$$

which is the desired identity matrix. The product matrix $\mathbf{E} = \mathbf{E}_2\mathbf{E}_1$ may be computed (noting carefully the sequence in which the matrices are listed) to give

$$\mathbf{E} = \mathbf{E}_2\mathbf{E}_1 = \begin{bmatrix} 3 & -1 \\ -2 & 1 \end{bmatrix}$$

Since $\mathbf{E}$ is a matrix such that $\mathbf{EA} = \mathbf{I}$, then $\mathbf{E}$ must be the matrix $\mathbf{B}$ we are looking for. Check for yourself that

$$\begin{bmatrix} 3 & -1 \\ -2 & 1 \end{bmatrix}\begin{bmatrix} 1 & 1 \\ 2 & 3 \end{bmatrix} = \begin{bmatrix} 1 & 0 \\ 0 & 1 \end{bmatrix}$$

The arithmetic has been set out above in a way which is helpful for the present exposition. For most purposes, however, it is more convenient to set it out as in Table 5.1. In Tableau 0, the given matrix is set out under the heading **A**, and an identity matrix of appropriate order is entered in the first part under the heading **I**. The first elementary operation is to subtract twice the first row from the second row; this is done with the *entire* first row of Tableau 0 to give the position in Tableau 1. Similarly, the second (and last) elementary

TABLE 5.1

	I		A	
Tableau 0	1	0	1	1
	0	1	2	3
Tableau 1	1	0	1	1
	−2	1	0	1
Tableau 2	3	−1	1	0
	−2	1	0	1

operation is to subtract the entire second row (in Tableau 1) from the first row, to give the position shown in Tableau 2. Thus the right-hand part of Tableau 1 has a record of $\mathbf{E}_1\mathbf{A}$, and that of Tableau 2 a record of $\mathbf{E}_2\mathbf{E}_1\mathbf{A}$. On the left-hand side, Tableau 1 records $\mathbf{E}_1$ and

Tableau 2 records $\mathbf{E}_2\mathbf{E}_1 = \mathbf{B}$; in other words, the elementary matrices are multiplied as the calculation proceeds. Check for yourself that the entries in the table bear this interpretation. This form of work-sheet is very convenient for pencil-and-paper calculation, and is also much the same record as would be kept inside an electronic computer if such a computer were used for a large calculation of this type.

This example shows a *constructive* approach to the task of establishing whether, for a given matrix **A**, there exists a matrix **B** such that $\mathbf{BA} = \mathbf{I}$. Whenever such a problem arises, we can tackle it by the use of elementary row operations in the manner shown in this section; if all turns out well (as it did in the above example), we are able to calculate the elements in the matrix, and hence show that it exists in the particular case studied.

5.3 *Exercises*

In each of the following cases, carry out elementary row operations on the given square matrix **A** in an attempt to find a matrix **B** such that $\mathbf{BA} = \mathbf{I}$. Where such a matrix **B** is found, check your arithmetic by premultiplying it into **A**; also compute **AB** in order to see whether **B** satisfies the full definition of the inverse of **A**. Where you cannot find a matrix **B**, study the arithmetical procedure in order to see why it has broken down.

1. $\begin{bmatrix} 2 & 1 \\ 1 & 1 \end{bmatrix}$

2. $\begin{bmatrix} 2 & 3 & 1 \\ 1 & 0 & 1 \\ 2 & 1 & 1 \end{bmatrix}$

*3. $\begin{bmatrix} 2 & 1 & 2 \\ 1 & 1 & 2 \\ 3 & 2 & 4 \end{bmatrix}$

5.4 *The existence of the inverse matrix*

This section establishes the conditions under which an inverse matrix exists, and extends the discussion of section 5.2 on how to calculate the inverse. To begin with, the same restriction is used as in section 5.2: we study the procedure for finding a matrix **B** such that when **B** is *premultiplied* into the given square matrix **A**, an identity matrix is obtained. (Later in this section, we consider whether the same matrix serves for *postmultiplication*, i.e. whether **AB** = **I**.)

The numerical cases given in the exercises in section 5.3 *suggest* that the procedure of applying elementary row operations succeeds in finding a matrix **B** such that **BA** = **I** if the echelon matrix discovered along the way has unit entries in *all* positions on the principal diagonal. The following theorem gives a more precise statement of the proposition.

Theorem 5.1. For a given square matrix **A** of order n, it is possible to find a sequence of elementary row operations, to be represented by the (square) matrix **B**, such that **BA** = **I**, if and only if **A** is non-singular.

Proof. Consider first the sufficient condition, i.e. the part of the theorem which says that a suitable sequence of elementary row operations can be found *if* **A** is non-singular. The first part of the sequence of elementary row operations leads to an echelon matrix, as indicated in the example in section 5.2. If **A** is non-singular, the echelon matrix is also non-singular (by Theorem 4.6) and hence it does not have any rows comprising zero entries (by Theorem 4.8). In other words, the echelon matrix has a unit entry in each position on the principal diagonal (cf. also example (2) in section 4.10). In this case it is always possible to complete the sequence in the manner illustrated in section 5.2; specifically, the unit entry in the last row and last column permits us to obtain zero entries everywhere else in the last column by subtracting multiples of the last row from all other rows; the unit entry in the second last row and column permits us to obtain zero entries in that part of the second last column which is above the principal diagonal, and so on until all entries above the principal diagonal have been made zero, and the echelon matrix is in turn transformed into an identity matrix. The entire sequence of elementary row operations then corresponds to the desired premultiplying matrix **B**.

Now consider the necessary condition, i.e. the part of the theorem which says a suitable sequence can be found *only if* **A** is non-singular. The only alternative form which **A** can take is that it is singular, and so this part of the proof will be argued by showing that if **A** is singular, it is not possible to find a suitable sequence of elementary row operations. It is immediately clear that the procedure just discussed (and illustrated in section 5.2) breaks down because the echelon matrix does not have a unit entry in each row. But this argument is not conclusive, because it *might* be possible to find a suitable sequence of elementary row operations by some other procedure. Fortunately, this possibility can be ruled out by the following argument: suppose that it is possible to find a suitable sequence, corresponding to a matrix **B** such that **BA** = **I**. Now **B** has rank n (by Theorem 4.7) while **A** is of rank less than n (since it is supposed here to be singular). Thus by Theorem 4.3, the rank of the product matrix **BA** must be less than n, which contradicts the equation since **BA** is equal to the identity matrix **I** of rank n. Thus the supposition (that a suitable sequence can be found) has led to a contradiction, and must therefore be false. Hence a suitable sequence can *not* be found if **A** is singular; in other words, a suitable sequence can be found *only if* **A** is non-singular.

Now that the condition has been established under which it is possible to find a matrix **B** such that **BA** = **I**, the time has come to consider whether the same matrix **B** will serve for postmultiplication. The exercises in section 5.3 suggest that it will, i.e. that **AB** = **I**. The general result is now stated and proved.

Theorem 5.2. Given a non-singular matrix **A** of order n, then the square matrix **B** which can be found such that **BA** = **I**, also satisfies the equation **AB** = **I**.

Proof. First note that since **A** and **B** are both square and of order n, then the product matrix **AB** is defined. Let

$$\mathbf{AB} = \mathbf{C}$$

where **C** is a square matrix of order n whose elements are not yet known. Postmultiply both sides of the equation by **A** to give

$$\mathbf{A(BA)} = \mathbf{CA}$$

Thus

$$\mathbf{AI} = \mathbf{CA}$$

and this may be written

$$\mathbf{IA} = \mathbf{CA}$$

Clearly, $\mathbf{C} = \mathbf{I}$ is a possible solution for $\mathbf{C}$, satisfying this equation. But $\mathbf{C}$ is *defined* as $\mathbf{AB}$ and hence must be unique. Thus $\mathbf{C} = \mathbf{I}$ is the only solution, and the theorem is proved.

These last two theorems may now be summarized together. It is possible to find a sequence of elementary row operations which transforms a given square matrix $\mathbf{A}$ into an identity matrix if and only if $\mathbf{A}$ is non-singular. This sequence may be represented by a matrix $\mathbf{B}$ which not only gives $\mathbf{BA} = \mathbf{I}$ but also $\mathbf{AB} = \mathbf{I}$. In other words the matrix $\mathbf{B}$ satisfies the full definition of an inverse matrix, as given in section 5.1. Thus the two theorems taken together suggest the following result:

Theorem 5.3. A given square matrix $\mathbf{A}$ has an inverse if and only if $\mathbf{A}$ is non-singular.

Proof. Theorem 5.1 shows that if $\mathbf{A}$ is non-singular, a certain matrix $\mathbf{B}$ can be found such that $\mathbf{BA} = \mathbf{I}$. Theorem 5.2 shows that $\mathbf{B}$ also satisfies $\mathbf{AB} = \mathbf{I}$, and hence $\mathbf{B}$ is an inverse matrix. This establishes the sufficiency part of the theorem.

Consider now the necessity of the condition, i.e. the proposition that an inverse exists *only if* $\mathbf{A}$ is non-singular. Theorem 5.1 shows that any procedure of applying elementary row operations works only if $\mathbf{A}$ is non-singular, but this still leaves doubt as to whether an appropriate matrix $\mathbf{B}$ could be found in some other way even if $\mathbf{A}$ is singular. The point is best tackled by a direct proof, as follows. Suppose that the inverse exists (and it may now be denoted $\mathbf{A}^{-1}$). Then

$$\mathbf{I}_n (= \mathbf{AA}^{-1})$$

can be regarded as a product matrix. Now the rank of $\mathbf{I}_n$ is n and hence, applying Theorem 4.3,

$$r(\mathbf{A}) \geqq n$$

Of course, since $\mathbf{A}$ is of order n, its rank cannot exceed n. Thus $\mathbf{A}$ is non-singular. In other words, an inevitable consequence of the existence of the inverse is that $\mathbf{A}$ is non-singular, i.e. the inverse exists *only if* $\mathbf{A}$ is non-singular.

Thus the overall result is very simple: an inverse exists if and only if **A** is non-singular. As already hinted, for any given numerical case, the approach is *not* to test first to see if **A** is non-singular, but rather just to start the sequence of elementary row operations in the manner indicated. Once the echelon matrix is reached, it is clear whether or not the matrix is non-singular; and if it is, the sequence can be completed to find the inverse.

We are now in a position to give a simple proof of Theorem 4.4. (This theorem has not been used so far in the derivation of other theorems, but it will later be so used.)

Theorem 4.4. If a matrix of rank k is premultiplied (or postmultiplied) by a non-singular matrix, the rank of the product matrix is k.

Proof. First consider premultiplication. Let **B** denote a matrix of order $m \times n$, having rank k (where, of course, $k \leqq \min [m, n]$). Let **A** denote the non-singular matrix; this is (by implication) a square matrix, since only square matrices are classified into singular and non-singular categories. Thus **A** must be of order $m \times m$ to permit the multiplication; hence **A** has rank m. Let the rank of **AB** be denoted by s.

From Theorem 4.3, $\quad s \leqq \min [m, k]$

Thus $\quad s \leqq k$, since $k \leqq m$

Now we may write $\quad \mathbf{B} = \mathbf{A}^{-1}(\mathbf{AB})$

since $\mathbf{A}^{-1}$ exists because **A** is non-singular.

Applying Theorem 4.3 to this equation (in which $\mathbf{A}^{-1}(\mathbf{AB})$ is regarded as the product of $\mathbf{A}^{-1}$ and **AB**), we obtain

$$k \leqq \min [r(\mathbf{A}^{-1}), s]$$

and, in particular, $k \leqq s$.

These two relationships between k and s lead to the conclusion that

$$s = k$$

For postmultiplication, the non-singular matrix which is postmultiplied into **B** must be square of order n and hence of rank n; the proof then follows precisely similar lines.

There is one other result to be established in this section. Some of the previous discussion has referred to *the* inverse matrix whereas it might have been prudent to allow for the possibility that **A** has more than one inverse. In fact, however, if an inverse exists, it is unique, as will now be shown:

Theorem 5.4. If a square matrix **A** has an inverse, then this inverse is unique.

Proof. Suppose that there exists a matrix **B** which satisfies the definition of an inverse, i.e. **B** is such that

$$\mathbf{AB} = \mathbf{BA} = \mathbf{I}$$

Suppose also that there exists another matrix **C** also such that

$$\mathbf{AC} = \mathbf{CA} = \mathbf{I}$$

(Note that **A**, **B** and **C** must all be square matrices of the same order, to ensure that all the multiplications are defined.) Consider $\mathbf{AB} = \mathbf{I}$ and premultiply by **C**, to give

$$(\mathbf{CA})\mathbf{B} = \mathbf{CI}$$

But $\mathbf{CA} = \mathbf{I}$, and this becomes

$$\mathbf{IB} = \mathbf{C}$$

or

$$\mathbf{B} = \mathbf{C}$$

Thus the two matrices are the same, and the inverse is unique.

Finally, it is worth noting that while elementary *row* operations have been used in the procedure for finding an inverse, there is the usual symmetry between rows and columns, and an equivalent procedure could be carried out by means of elementary *column* operations in order to find an inverse. This approach is not *exactly* similar but it is basically the same idea; it amounts to finding a matrix **F** such that $\mathbf{AF} = \mathbf{I}$. There are still other ways of computing inverse matrices. (Note however that it is not possible to use a *mixed sequence* of row and column operations in order to find the inverse matrix. It is always possible to find such a mixture to transform a non-singular matrix into an identity matrix; this is equivalent to finding square matrices **E** and **F** such that

$$\mathbf{EAF} = \mathbf{I}$$

But this is not comparable with the definition of an inverse matrix which is $\mathbf{AA}^{-1} = \mathbf{A}^{-1}\mathbf{A} = \mathbf{I}$; specifically we can *not* compare **EF** with $\mathbf{A}^{-1}$.)

5.5 *Some further properties of inverse matrices*

It is now time to establish a number of other widely used properties of inverse matrices.

Theorem 5.5. If **A** and **B** are non-singular matrices each of order n, then the product matrix **AB** is non-singular, and $(\mathbf{AB})^{-1} = \mathbf{B}^{-1}\mathbf{A}^{-1}$.

Proof. Since **A** and **B** are non-singular, then $\mathbf{A}^{-1}$ and $\mathbf{B}^{-1}$ exist, and are also of order n. Thus we may write

$$\begin{aligned}(\mathbf{B}^{-1}\mathbf{A}^{-1})\mathbf{AB} &= \mathbf{B}^{-1}(\mathbf{A}^{-1}\mathbf{A})\mathbf{B}\\ &= \mathbf{B}^{-1}\mathbf{I}_n\mathbf{B}\\ &= \mathbf{B}^{-1}\mathbf{B}\\ &= \mathbf{I}_n\end{aligned}$$

Also
$$\begin{aligned}(\mathbf{AB})(\mathbf{B}^{-1}\mathbf{A}^{-1}) &= \mathbf{A}(\mathbf{BB}^{-1})\mathbf{A}^{-1}\\ &= \mathbf{AI}_n\mathbf{A}^{-1}\\ &= \mathbf{AA}^{-1}\\ &= \mathbf{I}_n\end{aligned}$$

Thus, relative to the product matrix **AB** we have found a matrix $(\mathbf{B}^{-1}\mathbf{A}^{-1})$ which satisfies the full definition of the inverse matrix. In other words

$$(\mathbf{AB})^{-1} = \mathbf{B}^{-1}\mathbf{A}^{-1}$$

Also, since **AB** has an inverse, then it must be non-singular, by virtue of Theorem 5.3.

Note that the result can be generalized to the product of m non-singular matrices each of order n, the proof being similar to that employed in Theorem 3.3 which deals with the corresponding result for the transposition of matrices. Write out the required proof for yourself as an exercise.

Note also that in the above proof a matrix product of the general type $\mathbf{XAA}^{-1}\mathbf{Y}$ can be simplified and put equal to **XY** by noting that $\mathbf{AA}^{-1} = \mathbf{I}$. This technique is often of use. But it must be remembered that (as always) matrices are multiplied together in the order in which they are listed; the simplification noted above is valid because

$\mathbf{A}$ and $\mathbf{A}^{-1}$ are adjacent matrices in the sequence to be multiplied. If on the other hand the expression was $\mathbf{XAYA}^{-1}$ then since $\mathbf{A}$ and $\mathbf{A}^{-1}$ are no longer adjacent, such simplification is not possible.

Theorem 5.6. If $\mathbf{A}$ is a non-singular matrix, then $\mathbf{A}^{-1}$ is also non-singular and $(\mathbf{A}^{-1})^{-1} = \mathbf{A}$.

Proof. The inverse $\mathbf{A}^{-1}$ may be computed by applying a series of elementary row operations to $\mathbf{A}$. The matrix $\mathbf{E}$ representing this sequence is non-singular (by Theorem 4.7) and when it is multiplied into a non-singular matrix $\mathbf{A}$, then the product matrix $\mathbf{EA} = \mathbf{A}^{-1}$ must also be non-singular (by application of Theorem 4.4). Hence $\mathbf{A}^{-1}$ must itself have an inverse. By applying the definition of an inverse to the matrix $\mathbf{A}^{-1}$,

$$\mathbf{A}^{-1}(\mathbf{A}^{-1})^{-1} = \mathbf{I} = (\mathbf{A}^{-1})^{-1}\mathbf{A}^{-1}$$

Now $\mathbf{A}$ also has an inverse, and hence similarly

$$\mathbf{A}^{-1}\mathbf{A} = \mathbf{I} = \mathbf{AA}^{-1}$$

Comparing this pair of expressions and noting that the inverse of $\mathbf{A}^{-1}$ must be unique (by Theorem 5.4), we see that $(\mathbf{A}^{-1})^{-1}$ and $\mathbf{A}$ must be the same matrix.

Theorem 5.7. If $\mathbf{A}$ is a non-singular matrix, then $(\mathbf{A}')^{-1} = (\mathbf{A}^{-1})'$.

Proof. Since $\mathbf{A}$ is non-singular, $\mathbf{A}^{-1}$ exists and

$$\mathbf{AA}^{-1} = \mathbf{A}^{-1}\mathbf{A} = \mathbf{I}$$

Now take transposes (using Theorem 3.2)

$$(\mathbf{AA}^{-1})' = (\mathbf{A}^{-1})'\mathbf{A}' = \mathbf{A}'(\mathbf{A}^{-1})' = \mathbf{I}' = \mathbf{I}$$

This shows that $(\mathbf{A}^{-1})'$ is the inverse of $\mathbf{A}'$ since the definition of an inverse is satisfied. In other words

$$(\mathbf{A}')^{-1} = (\mathbf{A}^{-1})'$$

The matrix equation $\mathbf{AB} = \mathbf{0}$ does not necessarily imply that either $\mathbf{A} = \mathbf{0}$ or $\mathbf{B} = \mathbf{0}$. Two examples illustrating this have already been given – see example (3) of section 4.9 and example (6) of section 3.8. In both cases the matrices $\mathbf{A}$ and $\mathbf{B}$ are both singular. With the use of the inverse matrix concept, it is now possible to explore what happens if one of the matrices in the equation $\mathbf{AB} = \mathbf{0}$ is non-singular.

Theorem 5.8. If **A**, **B** are two matrices such that $\mathbf{AB} = \mathbf{0}$ and if **A** is non-singular, then **B** is a null matrix.

Proof. Note that **A** must be square if it is to be non-singular, but **B** need not be square.

Since **A** is non-singular, the inverse $\mathbf{A}^{-1}$ exists. Consider the equation

$$\mathbf{AB} = \mathbf{0}$$

and premultiply by $\mathbf{A}^{-1}$

to give
$$\mathbf{A}^{-1}\mathbf{AB} = \mathbf{0}$$
$$\therefore \quad \mathbf{IB} = \mathbf{0}$$
$$\therefore \quad \mathbf{B} = \mathbf{0}$$

One final point for this section should be noted. Since $\mathbf{I}_n\mathbf{I}_n = \mathbf{I}_n$ then $\mathbf{I}_n$ is its own inverse, or $\mathbf{I}_n^{-1} = \mathbf{I}_n$.

5.6 *Some computational aspects of finding the inverse*

It was remarked at the end of the previous chapter that to find the rank of a large matrix is not a trivial task, even when using an efficient computational approach such as the application of elementary row operations to derive the echelon matrix. In the present chapter, we have seen that when the given (square) matrix is non-singular, the elementary row operations may be continued after the echelon stage in order to find the inverse; clearly this involves even more computation than merely finding the rank, even though the use of elementary row operations is still a computationally efficient approach for the new task. Clearly, the amount of computation is greater, the larger is the matrix. Indeed, not only will the number of elementary operations increase as the number of rows in the matrix is larger, but each elementary row operation will involve more work because each row contains more elements. On these grounds, the amount of arithmetic increases more than in proportion to the order of the matrix. Indeed (as can be shown by a more thorough analysis) the amount is roughly proportional to n^4 where n is the order of the matrix; in other words the computational effort grows very rapidly indeed as we turn to larger matrices. A common view nowadays is that for n larger than about 5 or 6, it is best to use an electronic computer in order to find the inverse. If you think $n = 5$ is surprisingly small for this, test the matter for yourself by writing out a matrix of order 5 and carry out elementary row operations by hand, to seek an inverse. Of course, if n is only 5 or even (say) 10, finding the inverse is only a small job by the standards of an electronic computer. However, at the present stage of computer development, to invert a matrix of order 200 (say) is still quite a major operation.

The amount of work involved depends also on the degree of accuracy required. In each example given so far, the initial matrix **A** has always had integer elements and the arithmetic has been very simple, involving at worst a few simple fractions and always leading to an exact answer for $\mathbf{A}^{-1}$, exact in the sense that $\mathbf{A}^{-1}\mathbf{A}$ is exactly equal to an identity matrix. But in empirical studies, the elements of **A** are merely estimates (i.e. observed values) and even if the data are scaled so that these elements are all expressed as integers, the working usually cannot be kept in fractions; instead decimals must be used, and some rounding must be undertaken. As a consequence, the inverse will be approximate, as illustrated by the following example. Suppose that the matrix to be inverted is

$$\mathbf{A} = \begin{bmatrix} 1{\cdot}04 & 2{\cdot}39 \\ 2{\cdot}47 & 1{\cdot}83 \end{bmatrix}$$

Then the working may be set out in the usual way, as in Table 5.2. An echelon matrix is reached after three elementary row operations, and its form shows that the matrix does have an inverse. One further operation gives this inverse, shown at the bottom of the left-hand side of the table. Throughout, the working has been rounded to two decimal places. Check the arithmetic for yourself. The effect of this rounding on the accuracy of the computed inverse can be partially explored by multiplying this inverse into **A** itself, this time expressing the result to four decimal places:

$$\mathbf{A}^{-1}\mathbf{A} = \begin{bmatrix} 0{\cdot}9932 & -0{\cdot}0253 \\ 0{\cdot}0026 & 1{\cdot}0060 \end{bmatrix}$$

If this product matrix has its elements rounded to two decimal places, the result is

$$\mathbf{A}^{-1}\mathbf{A} = \begin{bmatrix} 0{\cdot}99 & -0{\cdot}03 \\ 0{\cdot}00 & 1{\cdot}01 \end{bmatrix}$$

and thus only one of its elements is 'correct' (i.e. the same as in $\mathbf{I}_2$), when measured to two decimal places. A greater degree of accuracy

TABLE 5.2

1	0	1·04	2·39
0	1	2·47	1·83
0·96	0·00	1·00	2·30
0·00	1·00	2·47	1·83
0·96	0·00	1·00	2·30
−2·37	1·00	0·00	−3·85
0·96	0·00	1·00	2·30
0·62	−0·26	0·00	1·00
−0·47	0·60	1·00	0·00
0·62	−0·26	0·00	1·00

could, of course, be achieved by carrying more decimal places throughout the calculation; the price of this is extra computational effort. As may be expected, the analysis of rounding errors soon becomes complicated, and goes beyond the level of this book.

The need to carry a considerable number of significant figures in the working in order to get reasonable accuracy in the inverse is, of course, an extra incentive encouraging the use of an electronic computer, which is in any case generally designed to work automatically with quite a large number of significant digits. Because matrix calculations are so commonly undertaken, and because computers are so useful in doing them, most computer installations are furnished with standard computer programs, commonly called 'matrix packages', which enable the user to have any desired sequence of matrix operations carried out on his own data. Since elementary row operations are an efficient approach to matrix inversion, the corresponding part of such a package generally uses them. However, in order to minimize rounding errors and to avoid certain other computational difficulties, the procedure is often a more sophisticated variant of the procedure given in this chapter.

5.7 *Exercises*

1. Calculate the inverse of

$$\mathbf{A} = \begin{bmatrix} 1 & 2 & 4 \\ 7 & 1 & 0 \\ 2 & 0 & 1 \end{bmatrix}$$

working in decimals and carrying two decimal places at each step of the calculation. Check the accuracy of your working by computing $\mathbf{AA}^{-1}$.

*2. Given the diagonal matrix $\mathbf{D} = [\lambda_i \delta_{ij}]$, under what circumstances does the inverse exist? Find the inverse when it does exist.

3. Use elementary *column* operations to show that

$$\mathbf{A} = \begin{bmatrix} 3 & 4 \\ 1 & 1 \end{bmatrix}$$

has an inverse, and then to compute the inverse.

5.8 *Computation of the inverse by using partitioning*

Under some circumstances it may be convenient to compute the inverse by exploiting the rules for the multiplication of partitioned matrices. In general, this permits the task of inverting a large matrix to be replaced by that of inverting two (or perhaps more) smaller matrices. As indicated at the beginning of section 5.6, the computational labour in finding an inverse increases more than proportionately with the order of the matrix, and hence this gambit can save considerable effort. In particular circumstances (illustrated below), the partitioning can exploit any special structure of the given matrix, which may make the computation even simpler. Finally, the partitioning technique is sometimes of analytical (as distinct from computational) interest, since it may permit an easier proof of some qualitative characteristic of the model being studied.

In order to study the basic idea of the partitioning approach, suppose that the non-singular matrix $\mathbf{A}$ (of order n) which is to be inverted, is partitioned into four submatrices, using the notation shown:

$$\mathbf{A} = \begin{bmatrix} \mathbf{E} & \mathbf{F} \\ \mathbf{G} & \mathbf{H} \end{bmatrix}$$

where $\qquad \mathbf{E}$ is of order $u \times u$

$\mathbf{H}$ is of order $v \times v$

$$\mathbf{F} \text{ is of order } u \times v$$

and $$\mathbf{G} \text{ is of order } v \times u$$

In other words, **E** and **H** are square matrices, while **F** and **G** are rectangular matrices. Note that $u+v=n$. Since **A** is non-singular, $\mathbf{A}^{-1}$ exists. Suppose that $\mathbf{A}^{-1}$ is partitioned in *exactly* the same way, using the following notation:

$$\mathbf{A}^{-1} = \begin{bmatrix} \mathbf{P} & \mathbf{Q} \\ \mathbf{R} & \mathbf{S} \end{bmatrix}$$

Now since $\mathbf{AA}^{-1} = \mathbf{I}_n$, we may use the rules developed in section 3.13 for the multiplication of partitioned matrices. Noting that the submatrices are conformable for multiplication in the following way, we may write:

$$\mathbf{AA}^{-1} = \begin{bmatrix} \mathbf{E} & \mathbf{F} \\ \mathbf{G} & \mathbf{H} \end{bmatrix} \begin{bmatrix} \mathbf{P} & \mathbf{Q} \\ \mathbf{R} & \mathbf{S} \end{bmatrix} = \begin{bmatrix} \mathbf{I}_u & \mathbf{0} \\ \mathbf{0} & \mathbf{I}_v \end{bmatrix}$$

where on the right-hand side, the two null matrices are of order $u \times v$, and $v \times u$ respectively; in other words $\mathbf{I}_n$ has been partitioned in *exactly* the same way as **A** and $\mathbf{A}^{-1}$. After working through the rules for this multiplication of partitioned matrices, the following four sets of equations are obtained:

$$\mathbf{EP} + \mathbf{FR} = \mathbf{I}_u \tag{5–1}$$
$$\mathbf{EQ} + \mathbf{FS} = \mathbf{0} \tag{5–2}$$
$$\mathbf{GP} + \mathbf{HR} = \mathbf{0} \tag{5–3}$$
$$\mathbf{GQ} + \mathbf{HS} = \mathbf{I}_v \tag{5–4}$$

where the null matrix in equations (5–2) is of order $u \times v$, and that in equations (5–3) is of order $v \times u$. Remember that the submatrices **E**, **F**, **G** and **H** are known, and that we wish to compute **P**, **Q**, **R** and **S** which comprise the inverse $\mathbf{A}^{-1}$.

In order to do this, we now solve these four *matrix* equations. From (5–3) we may write

$$\mathbf{HR} = -\mathbf{GP}$$

Now provided the given matrix and our scheme of partitioning is such that **H** is non-singular (which is *not* guaranteed by the fact that **A** is non-singular), then $\mathbf{H}^{-1}$ exists and we may premultiply by $\mathbf{H}^{-1}$ to give

$$\mathbf{R} = -\mathbf{H}^{-1}\mathbf{GP} \tag{5–5}$$

Now use this to substitute in (5–1) for **R**:

$$\mathbf{EP} - \mathbf{FH}^{-1}\mathbf{GP} = \mathbf{I}_u$$

or

$$(\mathbf{E} - \mathbf{FH}^{-1}\mathbf{G})\mathbf{P} = \mathbf{I}_u$$

Now the matrix $(\mathbf{E} - \mathbf{FH}^{-1}\mathbf{G})$ is square, and provided it is also non-singular, then its inverse exists; and we may premultiply by this inverse to give

$$\mathbf{P} = (\mathbf{E} - \mathbf{FH}^{-1}\mathbf{G})^{-1} \qquad (5\text{–}6)$$

Thus we can now use (5–6) to compute **P**, and then substitute in (5–5) for **P** in order to compute **R**. The other two submatrices may be calculated in similar fashion. From (5–4), we write

$$\mathbf{HS} = \mathbf{I}_v - \mathbf{GQ}$$

and premultiplying by $\mathbf{H}^{-1}$ gives

$$\mathbf{S} = \mathbf{H}^{-1}(\mathbf{I}_v - \mathbf{GQ}) \qquad (5\text{–}7)$$

Now substitute for **S** in (5–2) to yield

$$\mathbf{EQ} + \mathbf{FH}^{-1}(\mathbf{I}_v - \mathbf{GQ}) = \mathbf{0}$$

Thus

$$(\mathbf{E} - \mathbf{FH}^{-1}\mathbf{G})\mathbf{Q} = -\mathbf{FH}^{-1}$$

The matrix $(\mathbf{E} - \mathbf{FH}^{-1}\mathbf{G})^{-1}$ is simply **P** (from 5–6) and has already been calculated.

$$\mathbf{Q} = -\mathbf{PFH}^{-1} \qquad (5\text{–}8)$$

from which **Q** may now be calculated. Finally, **S** may be found from (5–7) after substituting from (5–8) for **Q**.

In summary, **P**, **R**, **Q** and **S** may be calculated in that sequence by using equations (5–6), (5–5), (5–8) and (5–7) respectively, provided only that **H** and $(\mathbf{E} - \mathbf{FH}^{-1}\mathbf{G})$ are non-singular.

Consider again the amount of computational effort involved. Instead of the direct computation of the inverse of **A** (which is of order n), the method requires the computation of the inverses of two smaller (square) matrices, namely **H** and $(\mathbf{E} - \mathbf{FH}^{-1}\mathbf{G})$ of order v and u (respectively) where $u + v = n$. (But in addition there is a certain amount of computation in the various matrix multiplications and subtractions.) Thus computational effort is saved when **A** is large; against this is the bother of organizing – or writing a computer program for – a somewhat more complex form of calculation.

A matrix may be partitioned into a larger number of parts. In the above case, if the partitioning is devised to make u small, then v may still be quite large. But the inversion of $\mathbf{H}$ (of order v) may be carried out by further partitioning, as a way of further substituting the inversion of a number of small matrices for the task of inverting one large matrix. More generally, the given matrix may be partitioned (in one operation, or in a sequence of steps) into any convenient number of submatrices.

The partitioning approach may be particularly valuable under two types of circumstance. First, $\mathbf{A}$ may have a special structure; for example it may be possible to partition so that $\mathbf{H}=\mathbf{I}_v$ in which case we know immediately that $\mathbf{H}^{-1}=\mathbf{I}_v$. Secondly, we may have already found the inverse of a matrix $\mathbf{H}$, and we may then wish to invert a larger matrix which includes $\mathbf{H}$ as a submatrix; this often arises in certain types of statistical calculation. These types of circumstance are illustrated by exercises in the next section.

Examples:

(1) Consider the matrix $\mathbf{A}$ partitioned as shown:

$$\mathbf{A}=\left[\begin{array}{c|cc} 2 & 3 & 1 \\ \hline 2 & 1 & 2 \\ 1 & 0 & 1 \end{array}\right]$$

Using the notation employed in this section, $\mathbf{H}=\begin{bmatrix}1 & 2\\ 0 & 1\end{bmatrix}$. This happens to be in echelon form as it stands, and $\mathbf{H}$ is clearly non-singular. Hence we know that $\mathbf{H}^{-1}$ exists. Further elementary row operations yield

$$\mathbf{H}^{-1}=\begin{bmatrix}1 & -2\\ 0 & 1\end{bmatrix}$$

(Check this for yourself.) Then

$$\begin{aligned}\mathbf{E}-\mathbf{F}\mathbf{H}^{-1}\mathbf{G} &= [2]-[3 \quad 1]\begin{bmatrix}1 & -2\\ 0 & 1\end{bmatrix}\begin{bmatrix}2\\ 1\end{bmatrix}\\ &=[1]\end{aligned}$$

This is just the identity matrix of order 1, and hence its inverse exists and is also [1]. Thus

$$\mathbf{P}=[1]$$

Then $\quad \mathbf{R}=-\mathbf{H}^{-1}\mathbf{G}\mathbf{P}=-\begin{bmatrix}1 & -2\\ 0 & 1\end{bmatrix}\begin{bmatrix}2\\ 1\end{bmatrix}[1]$

$$= \begin{bmatrix} 0 \\ -1 \end{bmatrix}$$

Next $\quad \mathbf{Q} = -\mathbf{PFH}^{-1}$

$$= -[1][3 \quad 1]\begin{bmatrix} 1 & -2 \\ 0 & 1 \end{bmatrix}$$

$$= [-3 \quad 5]$$

and $\quad \mathbf{S} = \mathbf{H}^{-1}(\mathbf{I}_2 - \mathbf{GQ})$

$$= \begin{bmatrix} 1 & -2 \\ 0 & 1 \end{bmatrix}\left\{\begin{bmatrix} 1 & 0 \\ 0 & 1 \end{bmatrix} - \begin{bmatrix} 2 \\ 1 \end{bmatrix}[-3 \quad 5]\right\}$$

$$= \begin{bmatrix} 1 & -2 \\ 3 & -4 \end{bmatrix}$$

Thus $\quad \mathbf{A}^{-1} = \begin{bmatrix} 1 & -3 & 5 \\ 0 & 1 & -2 \\ -1 & 3 & -4 \end{bmatrix}$

Check for yourself that $\mathbf{AA}^{-1} = \mathbf{I}_3$. Note that we do not know at the outset whether or not $\mathbf{A}^{-1}$ exists. If we find (as in this case) that $\mathbf{H}^{-1}$ exists and that $(\mathbf{E} - \mathbf{FH}^{-1}\mathbf{G})^{-1}$ exists, then this means that we can always construct $\mathbf{A}^{-1}$, and hence we have proved by construction that $\mathbf{A}^{-1}$ exists.

(2) Consider the following matrix, partitioned as shown:

$$\left[\begin{array}{cc:c} 1 & 2 & 3 \\ 1 & 0 & 1 \\ \hdashline 1 & 2 & 0 \end{array}\right]$$

Here $\mathbf{H} = [0]$ and clearly $\mathbf{H}^{-1}$ does not exist. But we are *not* entitled to conclude that the given matrix does not have an inverse. It may be simply that this particular partitioning does not work. But some other partitioning might be satisfactory. In the present case, let us try

$$\left[\begin{array}{c:cc} 1 & 2 & 3 \\ \hdashline 1 & 0 & 1 \\ 1 & 2 & 0 \end{array}\right]$$

Here $\mathbf{H} = \begin{bmatrix} 0 & 1 \\ 2 & 0 \end{bmatrix}$ and by elementary row operations we find that $\mathbf{H}^{-1}$ exists and is $\begin{bmatrix} 0 & \frac{1}{2} \\ 1 & 0 \end{bmatrix}$

$$\text{Then } \mathbf{E}-\mathbf{F}\mathbf{H}^{-1}\mathbf{G} = [1]-[2 \quad 3]\begin{bmatrix} 0 & \frac{1}{2} \\ 1 & 0 \end{bmatrix}\begin{bmatrix} 1 \\ 1 \end{bmatrix}$$
$$= [-3]$$

Thus this too has an inverse, and we can go on in the usual way to construct the complete inverse of the original matrix.

5.9 *Exercises*

*1. Compute the inverse of the following matrix by employing the partitioning which is shown:

$$\left[\begin{array}{cc|cc} 1 & 2 & 0 & 0 \\ 0 & 1 & 0 & 0 \\ \hline 3 & 4 & 1 & 0 \\ 1 & 2 & 0 & 1 \end{array}\right]$$

In what ways does the structure of the matrix simplify the computation?

*2. Consider the following matrix, partitioned as shown:

$$\left[\begin{array}{c|cc} 2 & 1 & 3 \\ \hline 1 & 1 & 2 \\ 1 & 0 & 1 \end{array}\right]$$

Note that this matrix is singular, since row 1 = row 2 + row 3. Thus the inverse does not exist. Explore what happens when the partitioning approach is used in an attempt to construct the inverse.

*3. Suppose that, in a statistical analysis of three variables, we have obtained a matrix which is the same as $\mathbf{A}$ in example (1) in section 5.8, and that $\mathbf{A}^{-1}$ has been computed as in that section. A fourth variable is then added into the analysis, leading to a need to compute the inverse of

$$\mathbf{B} = \begin{bmatrix} \theta & \mathbf{x}' \\ \mathbf{x} & \mathbf{A} \end{bmatrix}$$

where $\mathbf{x}' = [1 \quad 2 \quad 3]$ and $\theta = 0$. Use a partitioning scheme to compute $\mathbf{B}^{-1}$, given our knowledge of $\mathbf{A}^{-1}$.

*4. Given a non-singular matrix

$$\mathbf{A} = \begin{bmatrix} \mathbf{A}_1\, 0 & . & . & . & . & 0 \\ 0 \;\; \mathbf{A}_2 & & & & & . \\ . & & . & & & . \\ . & & & . & & . \\ . & & & & . & . \\ . & & & & \mathbf{A}_{n-1} & 0 \\ 0 & . & . & . & . \; 0 & \mathbf{A}_n \end{bmatrix}$$

where the submatrices $\mathbf{A}_1, \mathbf{A}_2, \ldots, \mathbf{A}_n$ are all square, find $\mathbf{A}^{-1}$.

5. (This exercise serves as an introduction to some of the problems studied in Chapter 6.) Suppose that part of an input–output model is as shown in the following table:

Value (in £000) of inter-sector sales

Producing sector	Consuming sector		Total sales
	A	B	
C	x_1	x_2	70
D	x_3	x_4	30
Total purchases	60	40	

The table is interpreted as follows: Sector A buys goods worth a total of £60,000 from sectors C and D, the separate values being x_1 (in £000) and x_3 (in £000) where these amounts are not known; similarly for sector D. Sector C sells a total of £70,000 worth to sectors A and B, but the components x_1 and x_2 (in £000) are not known; similarly for sector B.

Set up a system of equations in the form $\mathbf{Ax} = \mathbf{b}$. Is the matrix $\mathbf{A}$ singular or non-singular? Can the system be solved to obtain numerical values for the unknowns x_1 etc?

6. In the type of economic model which comprises a set of simultaneous linear equations, it is sometimes found desirable to re-formulate the original model by omitting one variable and one equation. In this context, suppose that the original coefficients matrix $\mathbf{A}$ (a square matrix of order n) has been inverted to yield

$\mathbf{A}^{-1}$, and that it is then decided to omit from $\mathbf{A}$ the last row and last column to yield a new coefficient matrix $\mathbf{B}$ which is supposed to be non-singular. Show how to compute $\mathbf{B}^{-1}$ given a knowledge of the elements of $\mathbf{A}^{-1}$ and regarding $\mathbf{A}^{-1}$ as being partitioned as

$$\mathbf{A}^{-1} = \begin{bmatrix} \mathbf{F} & \mathbf{g} \\ \mathbf{h}' & \beta \end{bmatrix}$$

where $\mathbf{F}$ is a square matrix, $\mathbf{g}$ and $\mathbf{h}'$ are column and row vectors respectively, and β is a scalar.

CHAPTER 6

The solution of simultaneous linear equations

6.1 *Introduction*

One of the most important and most common types of calculation in linear algebra is the solution of a system of simultaneous linear equations. We have already had occasion (in section 3.11 for example) to write out a general system of m equations involving n variables denoted x_j for $j = 1, \ldots, n$; such a system may be written out in full:

$$\begin{aligned} a_{11}x_1 + a_{12}x_2 + \ldots + a_{1n}x_n &= b_1 \\ a_{21}x_1 + a_{22}x_2 + \ldots + a_{2n}x_n &= b_2 \\ &\vdots \\ a_{m1}x_1 + a_{m2}x_2 + \ldots + a_{mn}x_n &= b_m \end{aligned}$$

If the vectors $\mathbf{x}$ and $\mathbf{b}$ and the matrix $\mathbf{A}$ are defined in the obvious way, then the system can be written:

$$\mathbf{Ax} = \mathbf{b}$$

In this chapter, we shall be concerned with whether or not such a system has a solution in any particular case, and (above all) with the computation necessary to find a solution.

By 'a solution' we mean a set of numerical values, one for each of the variables, such that *each* of the equations is satisfied by this set of values. It is for this reason that we speak of a system of *simultaneous* equations – one and the same set of values must simultaneously satisfy every equation in the system. As pointed out in section 1.1, an applied context may require the variables to be confined to non-negative values, and may even rule out all but integer values. In this chapter, it is supposed that the analysis may be conducted in terms of continuous variables (and that if necessary in an applied context, any fractional values may be rounded off at the end of the calculation). As far as the mathematical analysis is concerned, it is also supposed that in a solution to a system of equations, the variables may take on any values, positive or negative, integer or fractional; this is the supposition under which we will ascertain whether or not a system of equations has a solution. In applying this analysis to particular contexts, it will sometimes be necessary to rule out negative values; and even if a solution exists in the sense previously indicated, it does not necessarily follow that a non-negative solution exists. (The circumstances in which it is necessary to conduct the analysis in terms of integer variables right from the outset will be discussed briefly in Chapter 7 together with a very short introduction to some of the analytical problems which arise and some of the ways in which these problems can be tackled.)

Let us begin by considering a few simple examples of systems of simultaneous linear equations. The simplest system of all comprises one equation in one variable, i.e. with $m=n=1$. Such a system might be written

$$\alpha x=\beta$$

where α and β denote scalars whose values we know, and x denotes the single variable whose value we wish to find. In order to solve for x, we might be tempted to write out, without further thought, that

$$x=\beta/\alpha$$

But clearly, this makes sense (i.e. β/α is defined) only if $\alpha\neq 0$. Therefore let us distinguish two cases:

(i) If $\alpha\neq 0$, there is exactly one value of x which satisfies the system of equations, namely $x=\beta/\alpha$.

(ii) If $\alpha=0$, then we must enquire about the value of β before we can come to any conclusions:

(a) if $\beta\neq 0$, then the system has no solution; no matter what value x takes, the left-hand side αx is zero, while the right-hand side is non-zero; thus we cannot find a value for x which satisfies the equation

(b) if $\beta=0$, then there are an infinite number of solutions; in fact in this case *any* value of x satisfies the equation, since the left-hand side is always zero, and this time the right-hand side is zero.

The distinctions made in this very simple case turn up in much the same form in tackling the problem of solving a system of m equations in n variables. In particular we must find out whether there is any solution at all, and if so whether there is a range of alternative solutions. Further insight into these matters can now be obtained by looking at some examples which are a little more complicated. Let us consider some examples where there are two equations and two variables. First consider the system

$$\begin{aligned} 2x_1 + x_2 &= 4 \\ x_1 + 3x_2 &= 7 \end{aligned} \tag{6–1}$$

Let us attempt to find a solution by substitution. From the second equation write

$$x_1 = 7 - 3x_2$$

and substitute for x_1 in the first equation to yield

$$14 - 6x_2 + x_2 = 4$$

$$\therefore \quad x_2 = 2$$

and $$x_1 = 1$$

(Check for yourself that these values satisfy each of the two equations in (6–1).) Thus we have shown by construction that a solution exists; there were no complications and there does not appear to be any other solution.

For another example, consider

$$\begin{aligned} 2x_1 + x_2 &= 4 \\ 4x_1 + 2x_2 &= 7 \end{aligned} \tag{6–2}$$

The method of substitution breaks down and hence we cannot find a solution (at least by that method). Indeed, we observe that the left-hand side of the second equation is twice that of the first, but the same relation does *not* hold for the right-hand sides. Thus the equations are inconsistent; in other words, no values can be found for x_1 and x_2 which simultaneously satisfy both equations; any values which satisfy the first equation will *not* satisfy the second, and vice-versa.

For a third case, consider

$$\begin{aligned} 2x_1 + x_2 &= 4 \\ 4x_1 + 2x_2 &= 8 \end{aligned} \tag{6–3}$$

Here the left-hand sides have the same relationship as in equations (6–2) while the right-hand sides now also have this relationship. Thus in effect there is only one independent equation, in the sense that the second is satisfied by *all* values of x_1 and x_2 which satisfy the first, since the second is obtained from the first equation by multiplying by 2. Notice that the matrix of the coefficients

$$\begin{bmatrix} 2 & 1 \\ 4 & 2 \end{bmatrix}$$

has rank 1. In this case, we can now see that there are an infinite number of solutions, given by *all* the values of x_1 and x_2 which satisfy the first equation. Suppose we set $x_2 = \theta$ where θ can take on any numerical value, i.e. we regard θ as being what is called a parameter. Then we have $x_1 = 2 - \theta/2$, and thus our set of solutions is $(x_1, x_2) = ((2 - \theta/2), \theta)$ where θ can take on any numerical value whatsoever.

6.2 *Exercises*

1. For each of the systems (6–1), (6–2) and (6–3) draw a graph on which is entered the straight line corresponding to each equation of the pair. Hence give a geometrical interpretation of the results obtained in the text.
2. Explore the solution of the system

$$\begin{aligned} x_1 + 3x_2 &= 0 \\ 2x_1 + 6x_2 &= 0 \end{aligned}$$

Illustrate with a graph.

6.3 *A formal computational method*

We have seen that in examining a system of simultaneous linear equations, we must ascertain whether or not any solution exists, and if there is a solution we want to find the entire set of solutions where there is more than just a unique solution. In principle, we might want first to settle all questions of existence and uniqueness, and then go on to compute such solutions as do exist. In practical numerical work, however, we usually begin computation straight away, and in the course of it we check up on existence and uniqueness, at the same time as we work towards such numerical answers as can be obtained. (In theoretical contexts, on the other hand, we might simply want to establish whether or not a solution exists, without seeking answers in any specific numerical case.)

We have already used informal methods to compute solutions for very small-scale sets of equations. What is now needed is a more systematic approach which will give a computationally efficient way of finding the answers for large systems of equations. Such an approach will be introduced in this section in the context of a special case: suppose that the number of variables equals the number of equations. If the equations are denoted $\mathbf{Ax} = \mathbf{b}$, then on this supposition, $\mathbf{A}$ is a square matrix. For our special case, let us further suppose that $\mathbf{A}$ is non-singular. We can now quickly prove an important result for this case:

Theorem 6.1. If the matrix $\mathbf{A}$ is square and non-singular, the system of equations $\mathbf{Ax} = \mathbf{b}$ has a unique solution.

Proof. Since $\mathbf{A}$ is non-singular, $\mathbf{A}^{-1}$ exists. Consider the equations $\mathbf{Ax} = \mathbf{b}$ and premultiply both sides by $\mathbf{A}^{-1}$, to yield:

$$\mathbf{A}^{-1}\mathbf{Ax} = \mathbf{x} = \mathbf{A}^{-1}\mathbf{b}$$

Thus a solution exists. In order to show that it is unique, suppose that on the contrary there exists a second solution, denoted $\mathbf{x}^*$. Since this satisfies the equations, we have

$$\mathbf{Ax}^* = \mathbf{b}$$

as well as
$$\mathbf{Ax} = \mathbf{b}$$

where $\mathbf{x}$ here denotes the first solution. By subtraction,

$$\mathbf{A}(\mathbf{x} - \mathbf{x}^*) = \mathbf{0}$$

where the right-hand side is a vector of order n. Premultiply by $\mathbf{A}^{-1}$:

$$\mathbf{x} - \mathbf{x}^* = \mathbf{A}^{-1}\mathbf{0} = \mathbf{0}$$

Thus $\mathbf{x}^* = \mathbf{x}$ and the solution is unique.

Also note that our analysis so far has shown not merely that there is a solution and that it is unique, but also that we can find the solution by computing $\mathbf{A}^{-1}$ and premultiplying it into $\mathbf{b}$. This method of computation is basically equivalent to the formal method we are going to develop in this chapter for all systems of linear equations. In the present special case (where **A** is square and non-singular) this method could be applied as it stands. But in other cases, the method has to be adapted and extended, and to do this it is convenient to think of it in a somewhat different way. In preparation for this extension in later sections, the adapted method will be applied in this section to the special case presently being considered.

TABLE 6.1

	I		**A**	
Tableau 0	1	0	2	1
	0	1	1	3
Tableau 1	0·5	0	1	0·5
	0	1	1	3
Tableau 2	0·5	0	1	0·5
	−0·5	1	0	2·5
Tableau 3	0·5	0	1	0·5
	−0·2	0·4	0	1
Tableau 4	0·6	−0·2	1	0
	−0·2	0·4	0	1

In order to do this, consider the example of two variables and two equations given as equations (6–1) in the first section. As will be seen in a moment, this system has a non-singular matrix **A**, but we do not rely on this when we begin our computations. First let us compute the solution by the unadapted method, i.e. by finding the inverse $\mathbf{A}^{-1}$, and then premultiplying **b** by this inverse. The calculation of the inverse is set out in the usual way in Table 6.1; and we find that **A** is indeed non-singular, since we have no difficulties in transforming it into an identity matrix.

Finally
$$\begin{bmatrix} x_1 \\ x_2 \end{bmatrix} = \begin{bmatrix} 0{\cdot}6 & -0{\cdot}2 \\ -0{\cdot}2 & 0{\cdot}4 \end{bmatrix} \begin{bmatrix} 4 \\ 7 \end{bmatrix} = \begin{bmatrix} 1 \\ 2 \end{bmatrix}$$

Note that this agrees with our earlier solution of this system of equations.

In order to describe the adapted method, we need to define an *augmented matrix* in which the column vector **b** is added as a final extra column to the coefficients matrix **A**. Thus if the augmented matrix is denoted by **U**, then

$$\mathbf{U} = [\mathbf{A} \quad \mathbf{b}]$$

and in the present numerical example

$$\mathbf{U} = \begin{bmatrix} 2 & 1 & 4 \\ 1 & 3 & 7 \end{bmatrix}$$

TABLE 6.2

Tableau 0	2	1	4
	1	3	7
Tableau 1	1	0·5	2
	1	3	7
Tableau 2	1	0·5	2
	0	2·5	5
Tableau 3	1	0·5	2
	0	1	2
Tableau 4	1	0	1
	0	1	2

In the adapted method we carry out elementary row operations on the augmented matrix so as to obtain an identity matrix in the columns corresponding to **A**; the solution can then be read off from the final column in the transformed matrix. Thus the two separate phases (of finding $\mathbf{A}^{-1}$ and calculating $\mathbf{A}^{-1}\mathbf{b}$) are here combined into a single sequence of operations. Before justifying the logic of the adapted method, consider its application to the present example, as shown in Table 6.2. The **A** columns are transformed to an identity matrix by the same elementary row operations as before; the **b** column receives the same treatment and in the final tableau we can read off the solution $x_1 = 1$ and $x_2 = 2$. This method can be justified by observing that each elementary row operation carried out on the augmented matrix is equivalent to one of the steps carried out on the original equations when solving the equations by the elementary methods discussed in section 6.1. For example, in moving from Tableau 0 to Tableau 1, we multiply the first row of the augmented matrix by $\frac{1}{2}$, and this is equivalent to multiplying *both* sides of the first equation by $\frac{1}{2}$. To obtain Tableau 2, we subtract the new row 1 from row 2 and enter the result in row 2; this is equivalent to subtracting the new version of the first equation from the second equation. Thus each tableau corresponds to a set of equations which are derived from the initial set and which (by the rules of elementary algebra) have the same solution as the initial set. (Check this for yourself for each step of the calculation.) In particular Tableau 4 corresponds to

$$\begin{aligned} 1.x_1+0.x_2 &= 1 \\ 0.x_1+1.x_2 &= 2 \end{aligned}$$

and thus we can read off the solution in the manner already indicated. In effect then, our method is merely a systematic way of writing down the elementary algebraic procedures which we already know and which we used in section 6.1. However, as we shall see in later sections, by setting out the calculations in this way, we can use our knowledge of the rank of a matrix to help our understanding of the cases which are more complicated than the present one (where **A** is square and non-singular).

6.4 *Exercises*

*1. By using the computational method introduced in the previous section, find any solutions which exist for the following set of equations:

$$\begin{aligned} x_1+x_2+x_3 &= 6 \\ -x_1+2x_2+4x_3 &= 15 \\ 2x_1+2x_2+x_3 &= 9 \end{aligned}$$

*2. A firm may manufacture any or all of three products. Each unit of product I requires 1 unit of input 1, 2 units of input 2 and 1 unit of input 3. Each unit of product II requires 2 units of input 1 and 1 unit of input 3; the second type of input is not used in the manufacture of product II. For product III, the per unit requirements are 1, 1 and 2 units respectively of inputs 1, 2 and 3. The firm has available 7 units of input 1, 9 units of input 2 and 12 units of input 3. Find all possible production levels for the three products taken together which exactly use up all the available input supplies.

6.5 *Non-homogeneous equations: some general considerations*

In section 6.3, attention was confined to the system $\mathbf{Ax} = \mathbf{b}$ where $\mathbf{A}$ is square and non-singular. In this section we are going to examine a wider range of circumstances, namely those where $\mathbf{A}$ is any matrix of order $m \times n$, with m and n not necessarily equal; in other words we have m equations in n variables. One restriction will be imposed, however; we will assume that the vector $\mathbf{b}$ has at least one non-zero element, in which case the equations are called *non-homogeneous equations.* This term is used because an equation with a non-zero constant term on the right-hand side is a mixture of terms, some of which involve the variables raised to the power one (e.g. terms like $2x_2$) and one of which, the non-zero constant term, does not involve any variable. In contrast, a set of homogeneous equations may be written $\mathbf{Ax} = \mathbf{0}$ (where the right-hand side is a null vector); in each of these equations, all non-zero terms are similar in that each involves a variable raised to the power one.

Before obtaining any general results, let us consider an example of a set of non-homogeneous equations, namely the following system of three equations in two variables:

$$\begin{aligned} x_1 + 2x_2 &= 5 \\ 2x_1 + x_2 &= 4 \\ 2x_1 + 4x_2 &= 12 \end{aligned} \tag{6–4}$$

To this system we will apply *most* of the computational method developed in section 6.3; specifically we will form the augmented matrix, and then apply elementary row operations until we obtain the echelon matrix, as shown in Table 6.3. As in section 6.3, we can interpret each tableau in the calculation as representing a set of equations derived from the initial set, and having the same solution(s) – if any – as that initial set. Looking at Tableau 2 in particular, we see that the third row is equivalent to

$$0.x_1 + 0.x_2 = 1 \tag{6–5}$$

TABLE 6.3

Tableau 0			Tableau 1			Tableau 2		
1	2	5	1	2	5	1	2	5
2	1	4	0	−3	−6	0	1	2
2	4	12	0	0	2	0	0	1

Clearly *no* values for x_1 and x_2 would ever satisfy this equation. In other words, this derived set of equations has no solution. But this set is obtained from the initial set of equations by the usual elementary algebraic manipulation, just as in the example in section 6.3. In that section we argued that, for this reason, the derived set has the same solution as the initial set. In the present case then, since the derived set has no solution, then equally the initial set of equations has no solution. Notice that the elementary row operations carried out to obtain this third row in Tableau 2 are equivalent to the following operations: subtract twice the first equation of (6–4) from the third equation, and then divide the resulting equation by 2. Now the left-hand side of the third equation of (6–4) is twice that of the first equation, but the same does *not* hold for the *right*-hand sides of the two equations. In other words, these two equations are inconsistent, and this is the common-sense explanation of why the system has no solution.

Now consider whether we can learn any general lessons from this example. The echelon matrix in Tableau 2 of Table 6.3 is obtained from the augmented matrix **U**; this echelon matrix has rank 3, since there are three rows having non-zero entries (cf. Theorem 4.8). Next examine the matrix comprising the *first two* columns from Tableau 2. This is the echelon matrix obtained from the matrix **A**; clearly, its rank is 2. Also we know (from Theorem 4.6) that the rank of a matrix is the same as that of its echelon matrix. Thus we have $r(\mathbf{A}) = 2$ and $r(\mathbf{U}) = 3$, and we conclude that $r(\mathbf{A}) < r(\mathbf{U})$.

Now it is this discrepancy between the ranks which leads to the trouble, i.e. to the system of equations having no solution. If, in the general case, $r(\mathbf{A}) < r(\mathbf{U})$, then there will always be at least one equation like (6–5) in the set corresponding to the echelon matrix for

U; in other words, in the echelon matrix corresponding to **A**, there will always be at least one row of zero entries which does *not* correspond to a zero entry in the same row in the final column which is added on to give the echelon matrix for **U**. And hence in any such case, the system of equations has no solution.

This general argument could now be set out more formally and in greater detail in order to provide a full statement of the proof of the general proposition. Instead, however, the general theorem will simply be given a formal statement:

Theorem 6.2. If a system of non-homogeneous simultaneous linear equations $\mathbf{Ax}=\mathbf{b}$ is such that $r(\mathbf{A})<r(\mathbf{U})$, where **U** is the augmented matrix, then the system of equations has no solution.

This result prompts the question: under what circumstances (if any) can we guarantee that the system will have at least one solution? Again the result can be set out in terms of the ranks of **A** and **U**. First observe that $r(\mathbf{A})$ cannot exceed $r(\mathbf{U})$; this follows obviously by considering the echelon matrices obtained from **A** and **U** and applying Theorem 4.8 – the echelon matrix for **A** cannot have a larger number of rows with some non-zero entries than are to be found in the echelon matrix for **U** since **A** is merely part of **U**. Thus the only remaining class of circumstances we have to examine is when $r(\mathbf{A})=r(\mathbf{U})$.

Again, let us begin with an example, namely a set of equations which is derived from the set (6–4) by changing the right-hand side of the third equation:

$$\begin{aligned} x_1+2x_2&=5\\ 2x_1+x_2&=4\\ 2x_1+4x_2&=10 \end{aligned} \tag{6–6}$$

As before, form the augmented matrix and reduce it to the echelon form shown in Tableau 2 of Table 6.4. This time we see that the third row gives no trouble; there are zero entries throughout. (In common-sense terms, note that the third equation of (6–6) is simply twice the first equation, and we have removed the inconsistency which was present in the system (6–4).) Now we have $r(\mathbf{A})=r(\mathbf{U})=2$.

Also we can go on, as in Tableau 3, to subtract twice row 2 from row 1, to obtain a unit matrix of order 2 in the non-zero rows corresponding to **A** (i.e. looking only at the first two rows and first two columns in Tableau 3). We are now in a position to read off a solution, since the first two rows of Tableau 3 are equivalent to

$$1.x_1+0.x_2=1$$
$$0.x_1+1.x_2=2$$

TABLE 6.4

Tableau 0			Tableau 1			Tableau 2			Tableau 3		
1	2	5	1	2	5	1	2	5	1	0	1
2	1	4	0	−3	−6	0	1	2	0	1	2
2	4	10	0	0	0	0	0	0	0	0	0

Thus we find a solution $(x_1, x_2)=(1, 2)$. (In this example, we can speak of one of the equations being redundant, since there are only two independent equations, the third being a linear combination of the other two. Obviously any values for x_1 and x_2 which satisfy the first two equations will *automatically* satisfy the third equation, which therefore is redundant.) More generally, we can see that for *any* system for which $r(\mathbf{A})=r(\mathbf{U})$, the type of difficulty which we experienced with system (6–4) will not occur; we might therefore hope to go on to find a solution. This proposition is in fact justified; but we will prove it in a rather different manner:

Theorem 6.3. If a system of non-homogeneous simultaneous linear equations $\mathbf{Ax}=\mathbf{b}$ is such that $r(\mathbf{A})=r(\mathbf{U})$, where $\mathbf{U}$ is the augmented matrix, then there exists at least one solution to the system.

Proof. Suppose that the rank of $\mathbf{A}$ is k. In other words, suppose that

$$r(\mathbf{A})=r(\mathbf{U})=k$$

Note that $k \leqq n$, the number of variables. In section 4.8, we defined the rank of a matrix as the maximum number of linearly independent *rows* in the matrix, and argued that this is the same as the maximum number of linearly independent columns (cf. Theorem 4.2). Since $r(\mathbf{A})=k$, there must be at least one set of k linearly independent columns in $\mathbf{A}$. Let us suppose (without any loss of generality) that we have numbered the variables in such an order as to make the *first* k

columns of **A** linearly independent. Now this set of k columns also appears in the augmented matrix **U**. Since $r(\mathbf{U}) = k$, the maximum number of linearly independent columns in **U** is also k; hence *any* set of $k+1$ columns from **U** must be linearly dependent. In particular the set comprising the first k columns and the last column (which is **b**) must be linearly dependent. Let these first k columns from **U** (or from **A**) be denoted $\mathbf{a}_j$ (for $j = 1, 2, \ldots, k$). Since the set of $k+1$ columns is linearly dependent, then there must exist a set of scalars λ_j not all zero such that

$$\lambda_1 \mathbf{a}_1 + \lambda_2 \mathbf{a}_2 + \ldots + \lambda_k \mathbf{a}_k + \lambda_{k+1} \mathbf{b} = \mathbf{0}$$

by virtue of the definition of linear dependence. The question now to be settled is whether or not $\lambda_{k+1} = 0$. Suppose (for the moment) that it does. Then the last term in the above equation disappears, and there exists a set of scalars not all zero such that

$$\lambda_1 \mathbf{a}_1 + \lambda_2 \mathbf{a}_2 + \ldots + \lambda_k \mathbf{a}_k = \mathbf{0}$$

But these k vectors are linearly independent, which means that all these scalars $\lambda_1, \lambda_2, \ldots, \lambda_k$ must be zero. Thus the supposition (that $\lambda_{k+1} = 0$) has led to a contradiction, and so λ_{k+1} must be non-zero. Now divide the linear dependence equation above by $(-\lambda_{k+1})$ and transfer the term in **b** to the right-hand side:

$$-(\lambda_1 \mathbf{a}_1 + \lambda_2 \mathbf{a}_2 + \ldots + \lambda_k \mathbf{a}_k)/\lambda_{k+1} = \mathbf{b}$$

Now define a new set of scalars by the expression

$$x_j = -\lambda_j/\lambda_{k+1} \qquad (j = 1, 2, \ldots, k)$$

The previous vector equation may now be written

$$x_1 \mathbf{a}_1 + x_2 \mathbf{a}_2 + \ldots + x_k \mathbf{a}_k = \mathbf{b}$$

If this is written out fully in scalars, we obtain

$$\begin{array}{l} a_{11}x_1 + a_{12}x_2 + \ldots + a_{1k}x_k = b_1 \\ \cdot \\ \cdot \\ \cdot \\ \cdot \\ a_{m1}x_1 + a_{m2}x_2 + \ldots + a_{mk}x_k = b_m \end{array}$$

The argument so far may be summarized by saying that it is possible to find a set of k scalar values (here called x_j) which satisfy these m equations.

In other words, there must exist a solution to the equations $\mathbf{Ax} = \mathbf{b}$, in which solution, incidentally, $(n-k)$ of the variables are set equal to zero, while the others are permitted to take non-zero values. (In the present case, we supposed that the *first* k columns are linearly independent, and it is the corresponding variables which may be non-zero in the solution we have found. Nothing said here precludes the possibility that other solutions exist.)

Note that Theorem 6.2 may be proved by employing an argument closely similar to that just used in the proof of Theorem 6.3.

The results of this section may now be summarized. A set of non-homogeneous equations has at least one solution if and only if the rank of the augmented matrix **U** is the same as the rank of the coefficients matrix, **A**. As before, it must be emphasized that this result simply gives us a general insight into non-homogeneous equations. We do *not* attempt to classify the set according to the ranks of **A** and **U** before beginning the task of finding any solutions which exist. Instead we apply elementary operations broadly in the manner used in this section and in the course of our calculations we will discover the facts about the ranks, and then go on to find solutions where they exist. Some further details of these computations are studied in the next section.

6.6 *Exercises*

For each of the following systems of equations, find a solution if any exists.

*1.
$$\begin{aligned} 2x_1+x_2-x_3 &= 1\\ 4x_1+x_2+x_3 &= 9\\ 2x_1-x_2-x_3 &= -3\\ 4x_1+2x_2-2x_3 &= 0 \end{aligned}$$

*2.
$$\begin{aligned} x_1+x_2+x_3 &= 6\\ x_1-x_2-x_3 &= -4\\ x_1+x_2-x_3 &= 0\\ 2x_1 \qquad -2x_3 &= -4 \end{aligned}$$

6.7 *Non-homogeneous equations: further discussion*

In each of the examples worked out in the sections 6.3 and 6.5, the rank of **A** is the same as the number of variables. As suggested by the proof of Theorem 6.3, this relationship between $r(\mathbf{A}) = k$ and n, the number of variables, appears to be of some importance. We shall see in the present section that if the non-homogeneous equations have a solution, the relationship between k and n determines whether or not the solution is unique. At the same time we shall see how to extend the computational method based on elementary row operations, in order to find all solutions when there is indeed more than one.

We now embark on a complete review of all the types of case which can occur, including types already illustrated in earlier sections. Suppose then a set of m non-homogeneous equations in n variables, $\mathbf{Ax} = \mathbf{b}$, for which $r(\mathbf{A}) = r(\mathbf{U}) = k$. We know that at least one solution exists (by Theorem 6.3).

Suppose first that $k = n$. Here there are two cases, depending on the relationship between m and n. Note that k cannot exceed m, since the rank cannot exceed the number of rows in **A**. Since $k = n$ on our present supposition, then either $m = n$ or $m > n$. *If* $m = n = k$, then **A** is square and non-singular. Thus the solution is unique, by Theorem 6.1, and it may be found by carrying out the kind of computation illustrated in Table 6.2. *If on the other hand* $m > n$ *and hence* $k < m$, we have the kind of case illustrated by the system of equations (6–6). Since $k < m$, the echelon matrix derived from **U** will have $(m-k)$ rows comprising zero entries; in other words only k of the equations are independent of each other, and the other $(m-k)$ equations are redundant in the sense that they may be obtained from them. Once the echelon matrix has revealed this, the computation may be continued as illustrated in Table 6.4. In effect, we are left solving k equations in $n = k$ variables, the submatrix of order k is non-singular, and the solution is unique, for the same reasons as before. In the system (6–6), for example, we started with three equations in two variables; we found $k = 2$, i.e. only two equations are independent, the third being a multiple of the first; finally we solved for x_1 and x_2 by using the first two rows of Tableau 3 in Table 6.4.

That is all that needs to be said about the cases when $k = n$. *Now suppose that* $k < n$. Again there are two cases, depending on whether

$k=m$ or $k<m$; as before, k cannot exceed m. *Take first the case where* $k=m$. We shall begin with the following system (6–7) as an example; here $n=3$ and $m=2$, and in the course of the calculation we shall discover that $k=2$.

$$\begin{aligned} x_1+\ x_2+x_3 &= 6 \\ 2x_1+3x_2+x_3 &= 11 \end{aligned} \tag{6–7}$$

As before we form the augmented matrix, and derive its echelon matrix, shown in Tableau 1 of Table 6.5; from this echelon matrix it is clear that $r(\mathbf{A})=r(\mathbf{U})=2$. Thus the system has at least one solution (by Theorem 6.3). The computation can now be continued as follows: we focus attention on the columns of the echelon matrix which contain those unit entries which are the first non-zero entry in the row in question. In the present case there are two such entries, in the first row and first column, and in the second row and second column, of Tableau 1.

TABLE 6.5

Tableau 0				Tableau 1				Tableau 2			
1	1	1	6	1	1	1	6	1	0	2	7
2	3	1	11	0	1	−1	−1	0	1	−1	−1

If the second row is now subtracted from the first row, we obtain a zero in the second column and first row, as shown in Tableau 2. In other words, we carry out further elementary row operations to produce an identity matrix out of the columns containing these special unit entries in the echelon matrix; in the present example, we have unit entries in columns 1 and 2 only, in the echelon matrix, and in Tableau 2 we have obtained an identity matrix when we consider the first two columns only. (In other examples, the procedure may be a little more complicated; but the principle is the same.)

The final step of the calculation can best be understood by considering the equations which are equivalent to Tableau 2:

$$\begin{aligned} x_1 \qquad +2x_3 &= \ \ 7 \\ x_2-\ x_3 &= -1 \end{aligned}$$

(Check for yourself that – as usual – these equations can be derived from the original equations (6–7) by elementary operations on the equations themselves.) Note that the variables x_1 and x_2 each appear only once in the system; this corresponds to our identity matrix comprising columns 1 and 2 of Tableau 2. Thus we can transfer the remaining variable x_3 to the right-hand sides of the equations to make x_1 and x_2 the subjects:

$$\begin{aligned} x_1 &= 7-2x_3 \\ x_2 &= -1+x_3 \end{aligned}$$

In effect, we have now solved the system. In this form, the equations show that x_3 can take on any value and that we can then determine x_1 and x_2 accordingly. Thus we can speak of x_3 as a parameter. Let x_3 take on the value θ where θ is any number, positive, zero or negative. Then our solution is

$$\begin{bmatrix} x_1 \\ x_2 \\ x_3 \end{bmatrix} = \begin{bmatrix} 7-2\theta \\ -1+\theta \\ \theta \end{bmatrix} \tag{6–8}$$

For example, one solution occurs if $\theta=1$, which gives [5, 0, 1] as the solution vector; another occurs if $\theta=0$, yielding [7, −1, 0]. More generally we can check that this is indeed a solution for *any* value of θ by substituting these general values for x_1, x_2 and x_3 into each of the initial equations in (6–7). (For example, the left-hand side of the first equation becomes

$$\begin{aligned} x_1+x_2+x_3 &= 7-2\theta-1+\theta+\theta \\ &= 6 \end{aligned}$$

which is the same as the right-hand side.) Not only is the solution not unique, but since θ can take on an infinite number of different values, then there are an infinite number of solutions to the system (6–7).

As a further example, consider the same system of equations (6–7), but with the variables written down in a different order:

$$\begin{aligned} x_1+x_3+x_2 &= 6 \\ 2x_1+x_3+3x_2 &= 11 \end{aligned} \tag{6–9}$$

The computation may be carried out according to the same principles as before, as shown in Table 6.6. The echelon matrix is shown in

TABLE 6.6

Tableau 0				Tableau 1				Tableau 2			
1	1	1	6	1	1	1	6	1	0	2	5
2	1	3	11	0	1	−1	1	0	1	−1	1

Tableau 1, and from the final tableau, we can write out equations with x_1 and x_3 as subjects:

$$\begin{aligned} x_1 &= 5 - 2x_2 \\ x_3 &= 1 + x_2 \end{aligned} \tag{6–10}$$

If x_2 is set equal to a parametric value, denoted by ϕ this time, then the general solution vector is

$$\begin{bmatrix} x_1 \\ x_2 \\ x_3 \end{bmatrix} = \begin{bmatrix} 5-2\phi \\ \phi \\ 1+\phi \end{bmatrix}$$

At first sight, this may appear to differ from the previous statement in equation (6–8). But, in fact, the two statements are completely equivalent since any value of ϕ corresponds to a single value of θ, and vice-versa. As is clear from inspection, if we write $\phi = -1+\theta$, then (6–10) becomes the same as (6–8). To take a numerical example, if $\phi = 0$ and $\theta = 1$, then both statements give the same solution, namely [5, 0, 1]. Thus, though there are different ways of finding the infinite number of solutions, each method yields the same set of solutions.

In summary, this example is of two equations in three variables, and we found that $r(\mathbf{A}) = r(\mathbf{U}) = 2 = m < n$. We discovered that there are an infinite number of solutions, with one variable set equal to a parametric value. We also found that there was some choice in which variable could be chosen for this role; at first we used x_3, but in the second calculation we found it possible to use x_2. In each calculation, we see that any variable which does not correspond to the columns forming the identity matrix in the final tableau can be taken over to the right-hand sides of the equations and assigned parametric values. Since there are n variables and since the rank of this identity matrix is $r(\mathbf{A}) = r(\mathbf{U}) = k$, then in general $n-k$ variables can be treated in this way, and assigned parametric values. This leaves open the question of which of the n variables can be included in this set; this point is taken up below.

If desired, we could now set out a formal and general account of the particular type of calculation we have used for this example, in order to prove the general results. But the general line of argument follows very closely the working of the example, and so it will not be set out here. Instead we merely state, without general proof, the theorem which embodies the general results for this type of case:

Theorem 6.4. For a system of m non-homogeneous equations in n variables, if $r(\mathbf{A}) = r(\mathbf{U}) = k$ with $k = m < n$, then there exists an infinite number of solutions, and $n-k$ of the variables may be treated as parameters.

Before leaving this case, let us consider yet another example which illustrates some further complications which can arise:

$$\begin{aligned} x_1 + 2x_2 + x_3 + x_4 &= 8 \\ x_1 + 2x_2 - x_3 - 2x_4 &= 2 \end{aligned} \tag{6–11}$$

As before, carry out elementary row operations to obtain the echelon matrix shown in Tableau 2 of Table 6.7. This shows that $r(\mathbf{A}) = r(\mathbf{U}) = 2$, and hence the system has at least one solution (by Theorem 6.3). Since $n = 4$ and $r(\mathbf{A}) = k = 2$, then $n-k = 2$ of the variables can

TABLE 6.7

Tableau 0					Tableau 1				
1	2	1	1	8	1	2	1	1	8
1	2	−1	−2	2	0	0	−2	−3	−6
Tableau 2					**Tableau 3**				
1	2	1	1	8	1	2	0	−0·5	5
0	0	1	1·5	3	0	0	1	1·5	3

be assigned parametric values. At this stage, we again focus attention on the 'leading' unit entry in each row of the echelon matrix; this time such entries occur in the first and third columns, respectively. As before we carry out further elementary row operations to convert these columns into the constituents of an identity matrix, as shown in Tableau 3. And again, we can regard the other variables, here x_2 and x_4, as being transferred to the right-hand sides, so that Tableau 3 is equivalent to the system

$$\begin{aligned} x_1 &= 5-2x_2+0{\cdot}5\,x_4 \\ x_3 &= 3 \qquad\quad -1{\cdot}5\,x_4 \end{aligned}$$

Assign parametric values $x_2 = \theta$ and $x_4 = \phi$. Then the general solution vector is

$$\begin{bmatrix} x_1 \\ x_2 \\ x_3 \\ x_4 \end{bmatrix} = \begin{bmatrix} 5-2\theta+0{\cdot}5\phi \\ \theta \\ 3 \qquad -1{\cdot}5\phi \\ \phi \end{bmatrix} \tag{6–12}$$

(Again, check for yourself that this solution satisfies the system (6–11), by substituting these *general* values for the x_j into each of the equations.) Note that while the value of x_1 depends on both parameters, that for x_3 depends on ϕ only.

This example casts light on the question of which of the variables can be chosen for the assignment of parametric values. As we have seen, any one computation gives a unique choice, but if we write the variables down in a different order we may find that a new calculation leads to a different choice. The method of computation makes it clear that the remaining variables (i.e. those to which we do *not* assign parametric values) correspond to columns in the final tableau which together comprise an identity matrix of order k. (In our last example, this involved the first and third columns.) Similarly in the echelon matrix, these columns comprise a square submatrix of order k and of rank k. By applying Theorem 4.6 to this submatrix, we see that the corresponding columns of **A** must also form a submatrix of rank k. In other words, this set of variables (x_1 and x_3 in our last example) must correspond to linearly independent columns in the matrix **A**. Putting the point the other way round, any set of k variables which correspond to linearly independent columns in **A** would serve as the set to which parametric values are *not* assigned. Where more than one set exists, the set which is chosen by the computational procedure depends on the order in which the variables are listed. Notice that in this last example, x_1 and x_2 are the first two variables listed but these are not selected as the set in the computational procedure since the first two columns of **A** are a linearly *dependent* set of vectors, the second column being twice the first.

This completes the discussion of the case where $k = m < n$. *The final case of non-homogeneous equations occurs when* $k < n$ *and* $k < m$. To put the point loosely, in the previous case, we had some variables to spare (in the sense indicated in Theorem 6.4) and these could be assigned arbitrary parametric values; but $k = m$ and there were no redundant equations. In a still earlier case we had $k = n$ but $k < m$, and here there were some redundant equations; system (6–6) was an example of such a case (see section 6.5). In our final case, where $k < n$ and $k < m$, we combine these two characteristics; that is to say we have both redundant equations and spare variables. As before, let us begin with an example, which we will discover in due course to be a member of this class:

$$\begin{aligned} x_1 + x_2 + x_3 &= 4 \\ x_1 + 2x_2 - x_3 &= -1 \\ 2x_1 + 3x_2 \quad\;\; &= 3 \end{aligned} \tag{6–13}$$

The computation may be carried out in precisely the same manner as before, and is shown in Table 6.8. The echelon matrix is reached in

TABLE 6.8

Tableau 0				Tableau 1			
1	1	1	4	1	1	1	4
1	2	−1	−1	0	1	−2	−5
2	3	0	3	0	1	−2	−5
Tableau 2				**Tableau 3**			
1	1	1	4	1	0	3	9
0	1	−2	−5	0	1	−2	−5
0	0	0	0	0	0	0	0

Tableau 2 and this shows that $r(\mathbf{A}) = r(\mathbf{U})$; hence a solution exists (by Theorem 6.3).

We also note that $r(\mathbf{U}) = 2$, since the echelon matrix has one row of zero entries. In other words, the equations are not all independent of each other, and in this sense we have redundancy. (By inspection of the initial system (6–13), we see that the third equation is simply the sum of the first two equations.) This characteristic of our present example was also observed in the system (6–6). To continue the calculation we again focus attention on the leading unit entry in each row (having non-zero entries) in the echelon matrix; these entries occur in the first and second columns respectively. Further elementary operations transform these columns (ignoring the third, all-zero row) into an identity matrix, the entire transformed matrix being shown in Tableau 3. We can now read off the solution as before; the first two rows of this tableau are equivalent to

$$\begin{aligned} x_1 &= 9 - 3x_3 \\ x_2 &= -5 + 2x_3 \end{aligned}$$

If x_3 is set equal to θ a parameter, the general solution vector is

$$\begin{bmatrix} x_1 \\ x_2 \\ x_3 \end{bmatrix} = \begin{bmatrix} 9-3\theta \\ -5+2\theta \\ \theta \end{bmatrix}$$

(Again, check for yourself that this vector satisfies the initial system for any value of θ.)

As usual the constructive argument applied to this example could be set out in a more general way in order to prove the general result; here the theorem will simply be stated without setting out the proof:

Theorem 6.5. For a system of m non-homogeneous equations in n variables for which $r(\mathbf{A}) = r(\mathbf{U}) = k$ with $k < m$ and $k < n$, there exists an infinite number of solutions; $n - k$ of the variables may be assigned arbitrary parametric values, and $m - k$ of the equations are redundant.

As before, it must be remembered that the set of k variables which are not chosen for arbitrary assignment of parametric values must correspond to linearly independent columns in the matrix $\mathbf{A}$; also the k equations which are not redundant form a linearly independent set of vectors. In many examples, there will be some choice as to which variables may be regarded as the spare variables, and as to which equations may be regarded as redundant.

6.8 *Exercises*

For each of the following systems of equations, explore whether any solutions exist, and compute such solutions as can be found.

*1. $x_1 + x_2 + 2x_3 = 7$
$x_1 - x_2 + x_3 = 3$

2. $x_1 - x_2 + 3x_3 = 2$
$x_1 + x_2 + 2x_3 = 4$
$3x_1 - x_2 + 8x_3 = 8$

3. $2x_1 + x_2 + 4x_3 = 13$
$x_1 + 2x_2 + 3x_3 = 10$
$x_1 - x_2 + x_3 = 2$

4. $x_1 + 2x_2 + 3x_3 + 4x_4 = 29$
$x_1 - 2x_2 + x_3 + 2x_4 = 7$
$3x_1 - 2x_2 + 5x_3 + 8x_4 = 43$

*5. For the following system, do you notice any special property of the solution? What can be said about the columns of the coefficients matrix? Is there any connection between these two matters?

$$x_1 + x_2 - x_3 = -1$$
$$x_1 - x_2 + x_3 = 3$$

6.9 *Homogeneous equations*

Having completed the analysis of the various cases of non-homogeneous equations, we must now turn to homogeneous equations, where the general system may be represented $\mathbf{Ax} = \mathbf{0}$. As we shall see, the story is much the same as before although there are one or two important differences.

The first of these differences is that there *always* exists at least one solution for a system of homogeneous equations. By inspection, it is obvious that $\mathbf{x} = \mathbf{0}$ (in other words, a zero value for each of the variables) always makes the left-hand side $\mathbf{Ax}$ into a zero vector, and hence satisfies the system of equations. Such a solution is usually called the *trivial solution.* (Another way of seeing that there must always be a solution is to observe that the augmented matrix is obtained from $\mathbf{A}$ by adding a null column vector, and hence $r(\mathbf{A}) = r(\mathbf{U})$ always. Thus the echelon matrix derived from $\mathbf{A}$ will have the same number of rows composed exclusively of zero entries as will the echelon matrix derived from **U**. Hence there will never be any difficulty of the kind encountered in section 6.3 in dealing with system (6–4); rather, the calculations will, in this respect at least, be like those for system (6–6), and hence it will always be possible to find a solution. Notice that Theorems 6.2 and 6.3 can *not* be applied just as they stand, since each refers to non-homogeneous equations; but the argument underlying these theorems does apply.)

Clearly the next question is: under what circumstances (if any) do solutions exist other than the trivial solution? We shall shortly see that this depends on the relationship between k (the rank of **A**) and n, the number of variables. As usual k cannot exceed n, since the rank of a matrix cannot exceed the number of columns in it. Thus we have two cases to examine, according as $k=n$ or $k<n$. *Consider first the case where* $k=n$, and we begin with an example:

$$\begin{aligned} x_1+x_2&=0 \\ 2x_1-x_2&=0 \\ 4x_1+x_2&=0 \end{aligned} \tag{6–14}$$

Here we note that $m=3$ and $n=2$. We shall discover k in the course of the calculations. With homogeneous equations, there is no advantage in forming the augmented matrix, since the final column would be a null vector and we would not learn anything by including that column. Instead we carry out elementary row operations on the **A** matrix, the rationale being much the same as before. In the present case the echelon matrix is reached in Tableau 2 of Table 6.9, and from this we deduce that $r(\mathbf{A})=k=2$.

We continue, as usual, in order to obtain an identity matrix as in the first two rows of Tableau 3. Now consider the set of equations

TABLE 6.9

Tableau 0		Tableau 1		Tableau 2		Tableau 3	
1	1	1	1	1	1	1	0
2	−1	0	−3	0	1	0	1
4	1	0	−3	0	0	0	0

which are equivalent to these two rows (after including the zero entries on the right-hand sides):

$$1.x_1+0.x_2=0$$
$$0.x_1+1.x_2=0$$

These yield simply the trivial solution of which we already know. In those cases of non-homogeneous equations where we found an infinite number of solutions, the non-uniqueness of the solution arose because we were able to transfer certain variables to the right-hand sides of the equations, leaving on the left-hand sides those variables which corresponded to the columns of the identity matrix. In the present case, since $k=n$, all variables correspond to these columns, and so none can be transferred to the right-hand side. This suggests that the solution is unique, i.e. that no solution exists apart from the trivial solution. And this can be proved rigorously by an argument closely similar to that used to prove Theorem 6.1. Here, we merely state the general result, without going into the details of the proof.

Theorem 6.6. A system of linear equations, denoted $\mathbf{Ax}=\mathbf{0}$, and involving m homogeneous equations in n variables, has only the unique (trivial) solution $\mathbf{x}=\mathbf{0}$, if $r(\mathbf{A})=n$.

In such a case, note that the rank k can be equal to n only if $m \geqq n$, since the rank of a matrix cannot exceed the number of rows; in other words we are considering systems in which there are at least as many equations as there are variables.

Now consider *the second case in which* $k<n$. Again, begin with an example:

$$\begin{aligned} x_1+x_2+\ x_3 &= 0 \\ x_1-x_2-2x_3 &= 0 \end{aligned} \tag{6–15}$$

Here $m=2$ and $n=3$. As before, carry out elementary row operations on $\mathbf{A}$ until the echelon matrix is reached, as in Tableau 2 of Table 6.10. This shows that $r(\mathbf{A})=k=2$; thus $k<n$. In the columns having leading unit entries in the echelon matrix (i.e. in columns one and two), form an identity matrix by carrying out further elementary row operations on the matrix, as in Tableau 3. The equations which are equivalent to this tableau may now be written down, with x_3 moved over to the right-hand side:

$$\begin{aligned} x_1 \qquad &= \ \ 0{\cdot}5\, x_3 \\ x_2 &= -1{\cdot}5\, x_3 \end{aligned}$$

TABLE 6.10

Tableau 0			Tableau 1			Tableau 2			Tableau 3		
1	1	1	1	1	1	1	1	1	1	0	0·5
1	−1	−2	0	−2	−3	0	1	1·5	0	1	1·5

Thus, as usual, we may set $x_3 = \theta$, a parameter, and the general solution vector is

$$\begin{bmatrix} x_1 \\ x_2 \\ x_3 \end{bmatrix} = \begin{bmatrix} 0{\cdot}5\theta \\ -1{\cdot}5\theta \\ \theta \end{bmatrix} \tag{6–16}$$

In one important respect this kind of result is similar to that obtained for the corresponding case of non-homogeneous equations; but there are also some important differences. Let us deal first with the similarity: since $k < n$, then at least one variable can be assigned a parametric value, and hence there are an infinite number of solutions. Again, the argument used to demonstrate this for the present example could be set out more generally to prove the general result; but again the theorem will only be stated:

Theorem 6.7. A system of m linear homogeneous equations $\mathbf{Ax} = \mathbf{0}$ involving n variables has an infinite number of solutions if $r(\mathbf{A}) < n$.

Note that if $m < n$, then it is inevitable that $k < n$. In other words, if the number of equations is less than the number of variables, the system must have an infinite number of solutions, and not just the trivial solution $\boldsymbol{x} = \mathbf{0}$. Note also that, as with non-homogeneous equations, there may be redundant equations as well as spare variables.

We now turn to the differences between the present case and the corresponding case for non-homogeneous equations. First, notice that *all* the terms on the right-hand side of (6–16) involve the parameter θ; because the equations are homogeneous, there are no constant terms as there would be in the case of non-homogeneous equations – compare with (6–8), for example, where $x_1 = 7 - 2\theta$, an expression involving two terms, one a constant and one depending on the parameter θ. This result has a number of implications. First, it means that when the parameter or parameters are each set equal to zero, the solution becomes $\boldsymbol{x}=\mathbf{0}$. In other words, the set of solutions includes the trivial solution, as is to be expected. Also, in the present type of case where there is only one parameter, each member of the infinite set of solutions is a scalar multiple of each other member; for example, if in (6–16), $\theta=4$ and $\theta=2$ are considered, then the respective solutions are $(2, -6, 4)$ and $(1, -3, 2)$, and clearly the first solution is twice the second.

More generally, we see that if $\mathbf{x}^* \neq \mathbf{0}$ is a solution of $\mathbf{Ax}=\mathbf{0}$, then any scalar multiple $\lambda\mathbf{x}^*$ (where λ is any scalar) must be a solution, since $\lambda\mathbf{Ax}^* = \lambda\mathbf{0}$, and hence $\mathbf{A}(\lambda\mathbf{x}^*)=\mathbf{0}$. When $n-k=1$ (as in our example) there is only one parameter, and all solutions are scalar multiples of each other. However, when $n-k>1$ and there is more than one parameter in the general solution, then it is no longer true that all solutions have this relationship with each other; this is illustrated by exercise 5 in the next section. However it is still true, of course, that if $\mathbf{x}^* \neq \mathbf{0}$ is a solution, then $\lambda\mathbf{x}^*$ is also a solution.

In summary, we may say that all systems of homogeneous equations have at least the trivial solution, and that otherwise the story is much the same as for non-homogeneous equations; in particular, the solution is unique if and only if $k=n$, and there are an infinite number of solutions if and only if $k<n$. Our method for computing solutions is closely similar to that used for non-homogeneous equations.

6.10 *Exercises*

Find all solutions for each of the following systems of homogeneous equations.

1. $\begin{aligned} x_1+\ x_2&=0\\ 2x_1+2x_2&=0\end{aligned}$

2. $\begin{aligned} x_1+\ x_2+\ x_3&=0\\ x_1-\ x_2-2x_3&=0\\ x_1+2x_2-\ x_3&=0\end{aligned}$

3. $\begin{aligned} x_1+3x_2-3x_3&=0\\ x_1-\ x_2-\ x_3&=0\end{aligned}$

*4. $\begin{aligned} x_1+\ x_2-\ x_3&=0\\ x_1-\ x_2+\ x_3&=0\end{aligned}$

In this case, what do you notice about the form of the solution?

*5.
$$\begin{aligned} x_1 + x_2 + x_3 + x_4 &= 0 \\ x_1 - 3x_2 - x_3 - x_4 &= 0 \\ 4x_1 \quad\quad + 2x_3 + 2x_4 &= 0 \end{aligned}$$

In this case, you should find that two parameters are required. If these are denoted θ and ϕ respectively, evaluate the two solutions given by $\theta = 1$, $\phi = 0$ and $\theta = 0$, $\phi = 1$. What do you notice about the relationship between these two solutions?

6.11 *A summary of the results*

The previous sections have dealt in detail with a considerable number of different types of linear equation systems. Accordingly, it is convenient to give here a summary of the various different cases, as in Table 6.11. Remember that this table is only a summary, and that several other important points have to be borne in mind. Specifically, in all cases where the solution is not unique, the k variables which are not assigned arbitrary parametric values must correspond to a set of linearly independent columns in the matrix **A**. Similarly the k equations which are not redundant are linearly independent; any set of values which satisfies these k equations automatically satisfies the other $(m-k)$ equations. Also the relationship between m and n sometimes determines which case is relevant. For example, if $m < n$ and the equations are homogeneous, then $k < n$ and the system belongs to Case V; in other words, there are fewer equations than variables and hence automatically there are some 'spare' variables.

Much the same computational approach can be applied in all cases. Apply elementary row operations to the augmented matrix (for non-homogeneous systems) or to the **A** matrix (for homogeneous systems) until the echelon matrix is obtained. At this stage, it is then apparent which case applies. In Case I, no solution exists, and in Case IV only the trivial solution exists; no further calculation is necessary. In all other cases, continue with elementary row operations until the appropriate square submatrix (or in some cases the whole matrix) is transformed into an identity matrix. (The appropriate submatrix comprises the columns containing leading unit entries in the echelon matrix, ignoring any rows which have nothing but zero entries.) The general solution can then be written down by inspecting the final tableau, and assigning parametric values to those variables (if any) which correspond to columns not in the submatrix.

TABLE 6.11

Suppose m linear equations in n variables.

1. Non-homogeneous equations $\mathbf{Ax}=\mathbf{b}$ with $\mathbf{b}\neq\mathbf{0}$.
   Let $\mathbf{U}$ denote the augmented matrix $[\mathbf{A} \quad \mathbf{b}]$

Case I	$r(\mathbf{A})<r(\mathbf{U})$	No solution exists
Case II	$r(\mathbf{A})=r(\mathbf{U})=k=n$	Unique solution
Case III	$r(\mathbf{A})=r(\mathbf{U})=k<n$	Infinite number of solutions, with $(n-k)$ variables being assigned arbitrary parametric values

2. Homogeneous equations $\mathbf{Ax}=\mathbf{0}$. The trivial solution $\mathbf{x}=\mathbf{0}$ always exists.

Case IV	$r(\mathbf{A})=k=n$	Unique solution (the trivial solution)
Case V	$r(\mathbf{A})=k<n$	Infinite number of solutions, with $(n-k)$ variables being assigned arbitrary parametric values

Note: Cases II, III, IV and V each divide into two sub-cases according to the following scheme:

(a)	$k=m$	All equations are independent
(b)	$k<m$	$(m-k)$ of the equations are redundant

6.12 *A variant of the computational approach*

The computational method used in the previous sections and summarized in section 6.11 requires us to carry on with elementary row operations *after* the echelon matrix is found, so as to get to a position in which the solution can be read off by inspection. In the present section, we will describe a variant of this procedure; this variant requires a process of *back-substitution* to be used in the later stages of the calculation.

In order to study and compare the two procedures, let us consider the system

$$\begin{aligned} x_1+2x_2+6x_3 &= 16 \\ x_1+3x_2-\ x_3 &= 12 \\ x_1+2x_2 \qquad &= 10 \end{aligned} \tag{6–17}$$

TABLE 6.12

Tableau 0				Tableau 1			
1	2	6	16	1	2	6	16
1	3	−1	12	0	1	−7	−4
1	2	0	10	0	0	1	1
Tableau 2				**Tableau 3**			
1	2	0	10	1	0	0	4
0	1	0	3	0	1	0	3
0	0	1	1	0	0	1	1

First apply the original method: the echelon matrix is reached in Tableau 1 of Table 6.12, and since $r(\mathbf{A}) = r(\mathbf{U}) = 3$ for these non-homogeneous equations, then a solution exists (by Theorem 6.3). Further since $k = n = 3$, then the solution is unique; in other words the system belongs to Case II of Table 6.11. Further elementary row operations take us to Tableau 3 in Table 6.12. This tableau corresponds to the equations

$$\begin{aligned} x_1 \qquad\qquad &= 4 \\ x_2 \qquad &= 3 \\ x_3 &= 1 \end{aligned} \tag{6–18}$$

In other words, the solution can now be written down immediately, without any further computation.

The variant of this method begins with elementary row operations until the echelon matrix is reached, and then a process of back-substitution is applied. To see how it works, let us write down the equations which correspond to the echelon matrix in Tableau 1:

$$\begin{aligned} x_1 + 2x_2 + 6x_3 &= 16 \\ x_2 - 7x_3 &= -4 \\ x_3 &= 1 \end{aligned}$$

This gives the value of x_3 immediately. The next step is to substitute this value for x_3 (the last variable) into the second-last equation:

$$x_2 - 7 \times 1 = -4$$

$$\therefore \quad x_2 = 3$$

thus giving the value for x_2. Finally, the values for x_3 *and* x_2 are substituted into the first equation in order to obtain the value for x_1:

$$x_1 + 2 \times 3 + 6 \times 1 = 16$$

$$x_1 = 4$$

More generally, when the system of equations is such that the solution is not unique, or when some of the equations are redundant, the process of back-substitution can still be used. Of course, it is then applied to the singular submatrix which is formed from appropriate rows and columns of the entire matrix.

This process of back-substitution can save some computational effort compared with the original method, and in this sense the new method is more efficient. Both methods are among the more efficient ways of solving simultaneous equations, and many computer programs employ one or other method or some other closely related method.

6.13 *Exercises*

1. Use the back-substitution process to solve the following system of equations:

$$\begin{aligned} x_1 + x_2 + x_3 + x_4 + x_5 &= 15 \\ x_1 - x_2 + 3x_3 - x_4 + 3x_5 &= 23 \\ x_1 + x_2 + 2x_3 + 2x_4 + 2x_5 &= 24 \\ x_1 + x_2 - 4x_3 + x_4 + x_5 &= -5 \\ 3x_1 + x_2 + 6x_3 + 2x_4 + 6x_5 &= 62 \end{aligned}$$

2. Solve the following system of equations using the back-substitution process and carrying two decimal places at each step of the calculation. Check your results by substituting them in the original equations; carry four decimal places in this working, and report on any discrepancies you find.

$$\begin{aligned} 0{\cdot}34x_1 + 1{\cdot}32x_2 - 3{\cdot}79x_3 &= 0 \\ 1{\cdot}29x_1 - 0{\cdot}78x_2 + 1{\cdot}03x_3 &= 1{\cdot}21 \\ 0{\cdot}81x_1 - 0{\cdot}89x_2 &= 0{\cdot}74 \end{aligned}$$

*3. Suppose that a man is able to buy three kinds of food. Food I contains 1 unit of vitamin A and 1 unit of vitamin B per unit quantity of the food; each unit of Food II contains 2 units of vitamin A and 1 unit of vitamin B, while each unit of Food III contains 3 and 2 units of vitamins A and B respectively. The man wants to buy food so as to provide *exactly* 9 units of vitamin A and 7 units of vitamin B. What (if any) alternative purchasing programmes are open to him? (Suppose all units are divisible.)

*4. A wholesaler buys certain tonnages of three commodities. He then sells these amounts retail, his total receipts being £100, which includes a profit of £14·5. The retail and wholesale prices are as shown:

Commodity	A	B	C
Retail price (per ton)	£4	£8	£2
Wholesale price (per ton)	£3·5	£7	£1·5

To what extent (if at all) can we deduce the tonnages which he bought? Discuss your results, both in terms of the mathematical principles involved and in terms of their practical interpretation.

6.14 *Some applied examples*

This section is devoted to the description of some examples of the use of systems of simultaneous linear equations in social science contexts.

(1) In the linear model of a market for one commodity, given in exercise 4 of section 1.5, the equilibrium price was found by solving the system of three equations by an elementary approach. As an alternative, the more systematic method developed in this chapter can now be applied. The three equations may be written:

$$\begin{aligned} q_s \qquad\quad -bp &= -a \\ q_d \quad +dp &= c \\ q_s - q_d \qquad\quad &= 0 \end{aligned}$$

and from this, the augmented matrix may be written down in the usual way to give Tableau 0 of Table 6.13. After a few elementary row operations, an echelon matrix is reached in Tableau 2; note that since b and d are both positive, it is valid to divide by $(b+d)$. Since $r(\mathbf{A}) = r(\mathbf{U}) = 3 = n$, a unique solution exists; it may be obtained from Tableau 2 by back-substitution: from the third row $p = (a+c)/(b+d)$ and from the second row

$$\begin{aligned} q_d &= c - d(a+c)/(b+d) \\ &= (bc-ad)/(b+d) \end{aligned}$$

TABLE 6.13

Tableau 0				Tableau 1				Tableau 2			
1	0	$-b$	$-a$	1	0	$-b$	$-a$	1	0	$-b$	$-a$
0	1	d	c	0	1	d	c	0	1	d	c
1	-1	0	0	0	0	$b+d$	$a+c$	0	0	1	$(a+c)/(b+d)$

It is not necessary to use the first row to find q_s since in this case we know from one of the initial equations that $q_s = q_d$. For a small system such as this, the systematic approach is no quicker than the rudimentary method used in Chapter 1, although it does firmly establish whether or not the solution is unique. However, for a more general market model in which the quantities supplied and demanded of each of n commodities depend on all the n prices, the system of equations becomes much larger and more complex as n increases, and the systematic approach soon becomes highly advantageous. Exercise 1 in the next section deals with a two-commodity model of this kind.

(2) Consider again the input–output model introduced in sections 1.2 and 1.3. As has been seen in exercise 6 of section 3.15, the system of equations for the model for n industries may be written

$$\mathbf{x} = \mathbf{A}\mathbf{x} + \mathbf{b}$$

(where $\mathbf{A}$ is a square coefficients matrix of order n), and this in turn may be written

$$(\mathbf{I} - \mathbf{A})\mathbf{x} = \mathbf{b}$$

Of course, $(\mathbf{I} - \mathbf{A})$ is also a square matrix, and provided it is non-singular (which is to be expected in general in an input–output context), then its inverse exists; premultiply by this inverse to obtain

$$\mathbf{x} = (\mathbf{I} - \mathbf{A})^{-1}\,\mathbf{b}$$

which shows that one way of solving this system is to compute the inverse and multiply it into the vector $\mathbf{b}$. Alternatively, the augmented matrix may be transformed by elementary row operations. For the two-industry example given (and solved) in section 1.2, the working of this approach is set out in Table 6.14. The echelon matrix (in Tableau 2) has rank 2 and on the usual argument, there is a unique solution. From the second row of this tableau, $x_2 = 200$ and by back-substitution $x_1 = 350$. If just a single calculation is to be done, this method of setting it out is as good as any. But it may be desirable to explore such an input–output model for alternative numerical values for the elements of the final requirements vector, $\mathbf{b}$. In this case, it is easier to think of the solutions in terms of the previous equations, namely

$$\mathbf{x} = (\mathbf{I} - \mathbf{A})^{-1}\,\mathbf{b}$$

Once the inverse matrix has been computed, alternative solutions can be generated simply by multiplying this inverse into the alternative numerical values for the **b** vector.

TABLE 6.14

Tableau 0			Tableau 1			Tableau 2		
0·6	−0·8	50	1	$-1{\cdot}\dot{3}$	$83{\cdot}\dot{3}$	1	$-1{\cdot}\dot{3}$	$83{\cdot}\dot{3}$
−0·2	0·8	90	0	$0{\cdot}5\dot{3}$	$106{\cdot}\dot{6}$	0	1	200

(3) The next example deals with flows from one sector to another as time passes. The same basic ideas can be used in many contexts. To be concrete, we here speak of flows of people moving from one type of education to another, and in and out of the educational sector altogether, in a given country. We will treat time as a discrete variable and count the students and other people once every year. Let $\mathbf{x}_0$ be a column vector of order n, the elements of which measure the number of people in each sector at the beginning of the academic year 0; and similarly define $\mathbf{x}_1$, $\mathbf{x}_2$, etc. for later years 1, 2 etc. In a very simple example, we might distinguish four sectors: students in primary education; those in secondary education; those in other education; and those who are not students. Now define a square matrix **P** of order n, such that its typical element p_{rs} measures the proportion of the people in sector r at the beginning of one year who have moved to sector s by the beginning of the next year. For example, perhaps one-fifth (say) of those receiving primary education in one year will have moved into secondary education in the following year. Note that the matrix is not symmetric; for example, the proportion moving from secondary to primary education will be zero. The relation of the sector totals in one year to those of the following year is also affected by migration, births and deaths. We suppose that these numbers are not influenced by the size of the educational sectors and hence may be treated as constants; let $\mathbf{c}_0$ (a column vector of order n)

denote the net increase in each of the n sectors on this account at the end of year 0.

Then in sector s, number of people at beginning of year $1 = \sum_{r=1}^{n}$ (p_{rs} times number of people in sector r at beginning of year 0) + net increase in sector s on account of migration etc. In terms of our matrix notation, the set of such expressions (one for each sector) may be written

$$\mathbf{x}_1 = \mathbf{P}' \mathbf{x}_0 + \mathbf{c}_0$$

(Note that we write $\mathbf{P}'$ (the transpose of $\mathbf{P}$) since, in the previous expression, we needed to pick out p_{rs} for varying values of r; in other words, we use the proportions p_{rs} in a column of $\mathbf{P}$ and hence in a row of $\mathbf{P}'$.) This matrix equation can be regarded as a system in which $\mathbf{x}_0$ is known, the transition proportions $\mathbf{P}$ are estimated (on the basis of recent experience), the net increase vector $\mathbf{c}_0$ is similarly estimated, and the equation is used to predict $\mathbf{x}_1$, the set of sector totals for one year hence. (Note also that there is some formal similarity between this equation and the system $\mathbf{x} = \mathbf{A}\mathbf{x} + \mathbf{b}$ in the previous example of an input–output model.)

Let us now consider a numerical example of a four-sector model, having the sectors already mentioned. The matrix $\mathbf{P}$ is set out in Table 6.15; the figures assumed there are not unreasonable as a description of the situation in a developed country; the classification of the entire

TABLE 6.15

p_{rs}	1	2	3	4
	Students			Non-
	Primary	Secondary	Other	students
Students				
1. Primary	0·85	0·15	0	0
2. Secondary	0	0·80	0·03	0·17
3. Other	0	0	0·65	0·35
4. Non-students	0·02	0	0	0·98

population into only four sectors is too rough for practical purposes, but serves here to simplify the exposition. Note that in the matrix **P**, each row total is unity; this follows from the definition of p_{rs} as a *proportion* of those in sector *r*. The given sector populations (the elements of $\mathbf{x}_0$) are as shown in Table 6.16 and when the transition proportions are applied, the new vector $\mathbf{P}'\mathbf{x}_0$ is as shown in (the first four columns of) the next row of the table. Note that the total of this row is the same as the total of the previous row, because of the definition of p_{rs} as a proportion. (Check the arithmetic for yourself.)

TABLE 6.16

Numbers of people in each sector of the four-sector model (in thousands)

Sector	1	2	3	4	Total
$\mathbf{x}_0$	4,400	3,200	300	40,000	47,900
$\mathbf{P}'\mathbf{x}_0$	4,540	3,220	291	39,849	47,900
$\mathbf{c}_0$	30	10	9	151	200
$\mathbf{x}_1$	4,570	3,230	300	40,000	48,100

With the given data for $\mathbf{c}_0$ shown in the next row, the new sector totals (the elements of $\mathbf{x}_1$) are shown in the last row (in the first four columns) along with the new total population.

In the context of educational planning, the model may be used in more ambitious ways. For instance, repeated application of the matrix equation yields longer-range forecasts. Suppose that the matrix **P** is expected not to change over the next few years, and suppose that estimates are available of the net increase vectors $\mathbf{c}_1$, $\mathbf{c}_2$ etc. Then

$$\mathbf{x}_2 = \mathbf{P}'\mathbf{x}_1 + \mathbf{c}_1$$

and when this is combined with the previous equation, we have

$$\mathbf{x}_2 = \mathbf{P}'(\mathbf{P}'\mathbf{x}_0 + \mathbf{c}_0) + \mathbf{c}_1$$

Similarly, we can calculate $\mathbf{x}_3$ and so forth. The model can also be used to calculate backwards through time. Suppose that as part of a five-year development plan, a government sets a target figure for the educational sectors, represented by the elements of $\mathbf{x}_5$. Given assumptions about $\mathbf{P}$ and about the net increase vectors $\mathbf{c}_t$, it is then possible to compute the size of the educational sectors in the previous year, since

$$\mathbf{x}_5 = \mathbf{P}'\mathbf{x}_4 + \mathbf{c}_4$$

Hence

$$\mathbf{x}_4 = (\mathbf{P}')^{-1}(\mathbf{x}_5 - \mathbf{c}_4)$$

provided that $\mathbf{P}'$ has an inverse. Similarly $\mathbf{x}_3$ can be found in terms of $\mathbf{x}_4$, and so forth. Thus the government can see the position required in earlier years if the target is to be met. In summary, by setting out the model in matrix terms we can readily see what calculations we would make to explore any particular problem, and we also have an efficient way of organizing the calculations. (This type of model is also useful in other contexts where we wish to explore multi-sector populations – human or otherwise – and their changing pattern over time.)

6.15 *Exercises*

*1. Following the single-commodity market model introduced in exercise 4 of section 1.5 and discussed in the last section, suppose now that there are two inter-related markets, one for each of two commodities, and that the quantities supplied and demanded of each commodity each depend on p_1 and p_2, the prices in the markets for the two commodities. Specifically, if q_{d1} and q_{s1} are the demand and supply quantities for the first quantity, and q_{d2} and q_{s2} those for the second, let the demand and supply equations (assumed linear) be denoted:

$$q_{d1} = a_{01} + a_{11}p_1 + a_{21}p_2$$
$$q_{s1} = b_{01} + b_{11}p_1 + b_{21}p_2$$

$$q_{d2} = a_{02} + a_{12}p_1 + a_{22}p_2$$
$$q_{s2} = b_{02} + b_{12}p_1 + b_{22}p_2$$

where the a_{ij} and b_{ij} are coefficients with appropriate signs. In the equilibrium position, supply and demand are equal in each market:

$$q_{d1} = q_{s1}$$
$$q_{d2} = q_{s2}$$

Taking the six variables in the order q_{d1}, q_{s1}, q_{d2}, q_{s2}, p_1 and p_2, form an augmented matrix and transform it into an echelon matrix, stating any assumptions you make about the coefficients. Hence find the equilibrium value for p_2. (To avoid notational complexity, define new coefficients $c_{ij} = a_{ij} - b_{ij}$ whenever any expressions arise such as $a_{11} - b_{11}$.)

*2. In a large sample survey of U.K. university students, 3,140 respondents gave 'yes' or 'no' answers to each of the first two questions in the questionnaire. The completed questionnaires were analysed to yield the following information:

	No. of respondents answering	
	Yes	No
Question 1	1,738	1,402
Question 2	1,675	1,465

The team conducting the analysis now realizes that it needs to know how the answers to question 1 are related to those to question 2; in other words it wants to know how many students fell into each of four categories, answering the two questions respectively by yes/yes, yes/no, no/yes and no/no. One way to get this information is to sort all the questionnaires into four groups and do a direct count, but this involves a great deal of sorting effort. Devise a method which requires somewhat less sorting effort, and indicate in full how you would complete the calculations, giving your reasoning.

*3. (Revision exercise.) Given that $\mathbf{A} = \begin{bmatrix} 3 & 2 \\ 4 & \alpha \end{bmatrix}$ and $\mathbf{B} = \begin{bmatrix} 0{\cdot}5 & 2 \\ 0 & \beta \end{bmatrix}$

then

(i) find a set of necessary and sufficient conditions on the values of α and β which ensures that $\mathbf{A}+\mathbf{B} > \mathbf{AB}$

(ii) find a sufficient condition for α which is independent of the value of β, and which together with a necessary condition on β, ensures that $\mathbf{A}+\mathbf{B} > \mathbf{AB}$.

Prepare a graph to illustrate these conditions, and to show in particular why your sufficient condition is not a necessary condition.

*4. (Revision exercise). Prepare an algebraic description of the following problem; also draw a graph which illustrates the feasible region within which the farmer must plan his crop-planting:

A farmer has 100 acres of land which can be used for growing barley or oats. The yields are 50 bushels per acre of barley and 80 bushels per acre of oats. Labour requirements are 5 man-hours per cultivated acre, plus 0·2 man-hours per bushel of barley and 0·0625 man-hours per bushel of oats. The farmer has only 1,200 man-hours of labour at his disposal, and in addition he expects he will be unable to arrange to transport more than 6,400 bushels in total from the farm to the merchant, to whom he sells the barley at £0·8 per bushel and the oats at £0·9 per bushel. He wishes to maximize his total gross receipts from the merchant.

*5. An employer has a staff which is divided into three grades. He has a well-established promotion policy which he proposes to maintain in the future. It may be described in terms of the matrix

$$\mathbf{P} = \begin{bmatrix} 0{\cdot}60 & 0{\cdot}30 & 0 \\ 0 & 0{\cdot}85 & 0{\cdot}10 \\ 0 & 0 & 0{\cdot}90 \end{bmatrix}$$

where p_{ij} measures the proportion of staff in the ith grade who are in the jth grade one year later. Note that the elements in a row of the matrix sum to less than unity, in the first row, for example, the sum is 0·9. The remaining 10% of the first grade leave the employer during the year (as a result of resignation, retirement or death); it may be supposed that the proportions who leave will remain constant for the next few years. The employer has 500, 2,000 and 600 staff in the three grades respectively. In two years' time he wants these totals to become 500, 2,100 and 900 respectively. In addition to making promotions as usual, he is prepared to recruit new staff for direct entry to each of the three grades, and he would like the number recruited to a grade to be the same in the second year as in the first year. Can he achieve all these aims, and if so, how many recruits should he take each year in each grade? (Hint: before carrying out any arithmetic, formulate the problem in general matrix terms and carry out all necessary manipulation of the general matrix equations in order to discover what arithmetical calculations are appropriate; suppose that all staff changes take place at discrete intervals of one year.)

CHAPTER 7

Integer variables and other topics

7.1 *Introduction*

This chapter comprises four parts:

basic solutions	sections 7.2 and 7.3
linear programming	sections 7.4 and 7.5
quadratic forms	sections 7.6 and 7.7
integer variables	sections 7.8 to 7.15

Of these, the last part on systems involving integer variables is the most important. Each part is independent of the other parts except that (a) some of the section on linear programming requires a knowledge of basic solutions, and (b) some use is made of linear programming ideas in the later part of section 7.13 and in the following sections.

7.2 *Basic solutions for linear equations*

This section is concerned with a certain kind of solution which can be found for a system of m simultaneous linear equations in n variables, which may be written $\mathbf{Ax}=\mathbf{b}$ where $\mathbf{A}$ is of order $m \times n$, and where it is supposed that $m<n$. As has already been remarked (in section 6.11), when there are fewer equations than variables, then $r(\mathbf{A})<n$ and there are some variables 'to spare'. More precisely, if $r(\mathbf{A})=k$, then any set of $(n-k)$ of the variables may be assigned parametric values provided only that the remaining k variables correspond to a set of k linearly independent columns of $\mathbf{A}$. In this section we are interested in the case where $r(\mathbf{A})=k=m$; in other words we suppose that none of the equations is redundant. And the kind of solution of interest is that obtained by setting each of the parameters equal to zero.

Example:

(1) Consider the set of two equations in three variables:

$$\begin{aligned} x_1+2x_2-\ x_3 &= 4 \\ x_1-\ x_2+2x_3 &= 1 \end{aligned} \qquad (7\text{–}1)$$

After forming the augmented matrix and applying elementary row operations in the usual way, the following tableau is obtained

$$\begin{matrix} 1 & 2 & -1 & 4 \\ 0 & 1 & -1 & 1 \end{matrix}$$

This shows that x_3 (corresponding to the third columns) may be assigned a parametric value. In particular, let us set $x_3=0$; by back-substitution, it is immediately found that $x_1=2$ and $x_2=1$. This is an example of the kind of solution in which we are interested. The system has $n=3$ and $m=2$; thus $n-m=1$, and one variable is set equal to zero; values are then found for the other two variables (which, as the tableau shows, correspond to linearly independent columns of the coefficients matrix). This kind of solution can now be given a formal definition.

Definition. Given a system of simultaneous linear equations $\mathbf{Ax}=\mathbf{b}$ with $\mathbf{A}$ of order $m\times n$ and $m<n$, and with $r(\mathbf{A})=m$, and if any non-singular submatrix of order m is chosen from $\mathbf{A}$, then the solution for $\mathbf{x}$ obtained by setting to zero the other $(n-m)$ variables and solving for the m variables associated with the submatrix is called a *basic solution.*

In this context of m independent equations, a basic solution thus has no more than m variables with non-zero values. As will be seen below, not all these m variables necessarily turn out to be non-zero. But the m variables which *can* be non-zero are called the *basic variables*. Note that once the set of m variables has been chosen, the values for them are uniquely determined; this follows (as in the above example) because of the nature of the echelon matrix for the m columns (which are of course linearly independent). The term '*basic* solution' is used because the m variables correspond in the matrix $\mathbf{A}$ to m linearly independent columns in E^m. Thus the columns form a basis (cf. the definition of a basis, and Theorem 2.5 in section 2.7).

Let us next consider how many different basic solutions can be found for a given system of equations. For each basic solution, it is a matter of selecting m variables out of the set of n. If this were the only consideration (which it is not), it would be easy to find out how many basic solutions there are. For example, for the system (7–1), there are three ways of choosing two variables; these are x_1 and x_2 (giving the basic solution already found); x_1 and x_3; and x_2 and x_3. In fact, for this example, all three selections work satisfactorily; the other two basic solutions are, respectively, $[3 \quad 0 \quad -1]$ and $[0 \quad 3 \quad 2]$; these may be found by listing the variables in the orders x_1, x_3, x_2 and x_2, x_3, x_1 respectively, and then applying elementary row operations in the usual way; check this for yourself. But, of course, this is not the whole story. The method of solving the equations works only if the selected set of m variables corresponds to m linearly independent columns in the **A** matrix, and this qualification is incorporated in the definition of a basic solution. In the system (7–1) every *pair* of columns from **A** is linearly independent (as may be checked easily) and this is why every one of the three possible selections for the set of $m(=2)$ variables leads to a basic solution. The next example illustrates how the lack of this linear independence cuts down on the number of basic solutions.

Examples:

(2) Consider the system

$$\text{(7–2)} \qquad \begin{aligned} x_1+2x_2 \phantom{{}+x_3} &= 1 \\ x_1+2x_2+x_3 &= 3 \end{aligned}$$

When elementary row operations are applied to the augmented matrix, the following tableau is obtained:

$$\begin{matrix} 1 & 2 & 0 & 1 \\ 0 & 0 & 1 & 2 \end{matrix}$$

Now consider the three ways in which two variables may be selected:

(i) x_1 and x_2: The corresponding columns in the tableau do not give an echelon matrix of rank 2, and hence (cf. section 6.7) we can *not* go on to get a solution by assigning a parametric value to the other variable. Thus x_1 and x_2 do not lead to a basic solution. This follows because the corresponding columns of **A** are $\begin{bmatrix}1\\1\end{bmatrix}$ and $\begin{bmatrix}2\\2\end{bmatrix}$ which clearly are linearly dependent.

(ii) x_1 and x_3: These correspond to a satisfactory echelon matrix in the tableau and hence may be used to give the basic solution [1 0 2]. The corresponding columns of **A** are linearly independent.

(iii) x_2 and x_3: If the above tableau has its first row divided by 2, the columns corresponding to these two variables form a satisfactory echelon matrix; this leads to the basic solution [0 $\frac{1}{2}$ 2]. The corresponding columns of **A** are linearly independent.

(3) The next example shows that in a basic solution, some of the m selected basic variables may themselves take on zero values. Consider the system

$$\begin{aligned} x_1 \quad\;\; + 2x_3 &= 2 \\ 2x_1 + x_2 - \;x_3 &= 4 \end{aligned} \tag{7–3}$$

Apply elementary row operations to the augmented matrix to obtain the tableau

$$\begin{array}{cccc} 1 & 0 & 2 & 2 \\ 0 & 1 & -5 & 0 \end{array}$$

Since the first two columns are linearly independent and hence the echelon matrix comprising these columns has rank 2, then x_3 may be assigned a zero value, leaving x_1 and x_2 as the basic variables. The basic solution is [2 0 0]; in other words only one of the two basic variables is non-zero. As may readily be calculated, the other two basic solutions are:

[2 0 0] if the basic variables are x_1 and x_3
[0 5 1] if the basic variables are x_2 and x_3

Thus in one case, we get the same numerical values as before, and in the other case, both basic variables are non-zero.

Definition. A basic solution is said to be *degenerate* if one or more of the basic variables takes on a zero value.

As the last example illustrates, degeneracy arises if and only if the vector **b** and *any* set of $(m-1)$ columns from the basic set are a linearly dependent set. For example, in the first basic solution found for the system (7–3), the basic columns from **A** are

$$\begin{bmatrix}1\\2\end{bmatrix} \text{ and } \begin{bmatrix}0\\1\end{bmatrix} \text{ while } \mathbf{b} = \begin{bmatrix}2\\4\end{bmatrix}$$

Although **b** and the second column from **A** are linearly independent, **b** and the first column from **A** are linearly dependent (since one is twice the other) and this is why the corresponding basic solution is degenerate, as may be seen by considering the elementary row operations which lead to the tableau: because of the linear dependence, the operations which secure a zero in the second row of the first column also yield a zero in the second row of the last column, and it is this zero entry which makes $x_2 = 0$ in the basic solution.

7.3 *Exercises*

*1. Find all basic solutions for the following system, note which of them are degenerate, and relate the results to the columns of the augmented matrix:

$$\begin{aligned} x_1 + 2x_2 + x_3 + 2x_4 &= 6 \\ 2x_1 + 4x_2 \qquad + x_4 &= 3 \end{aligned}$$

*2. In the spirit of the example of the baker (given in section 1.8), suppose that a manufacturer is able to produce amounts x_1 and x_2 of two products, provided he keeps within a feasible region given by

$$\begin{aligned} x_1 + 2x_2 &\leqq 4 \\ 2x_1 + x_2 &\leqq 6 \end{aligned}$$

These two constraints arise out of limited supplies of commodities called A and B respectively. As usual we require x_1 and x_2 to be non-negative. Prepare a graph showing this feasible region. Now define

$$\begin{aligned} x_3 &= \text{number of units of A not used up} \\ x_4 &= \text{number of units of B not used up} \end{aligned}$$

The two constraints may now be written as *equalities*:

$$\begin{aligned} x_1 + 2x_2 + x_3 \qquad &= 4 \\ 2x_1 + x_2 \qquad + x_4 &= 6 \end{aligned}$$

(Each equality says: total allocation of commodity, including amount not used, equals total amount available.) Find all basic solutions for this pair of equations. Now remember that x_1 and x_2 must be non-negative; also, because of the way they are defined, x_3 and x_4 must be non-negative. Ignoring basic solutions which include negative elements, associate each other basic solution with a point on your graph (i.e. for such a basic solution, find the point on the graph which has coordinates equal to the values for x_1 and x_2 in the basic solution).

*3. Find all basic solutions for the following system, and relate the results to the characteristics of the columns of the augmented matrix:

$$\begin{aligned} x_1 + 2x_2 + x_3 + x_4 &= 3 \\ x_2 + 2x_3 + 2x_4 &= 1 \\ x_1 \quad\quad + x_3 \quad\quad &= 1 \end{aligned}$$

7.4 *An appreciation of linear programming*

Although developed as recently as the 1940s, linear programming has rapidly become one of the most widely worshipped gods of the twentieth century. Notwithstanding this uncritical adulation and the many successful attempts to misapply the technique, linear programming *is* a very useful computational procedure which enables us to 'solve' (i.e. gain a great deal of extra understanding of) many policy problems requiring quantitative decisions, both in business and government, and in other social science contexts. To introduce the subject, let us begin with an example similar to that of the baker first discussed in section 1.8. In that example, the baker had limited supplies of *three* ingredients. To simplify the problem a little, let us now suppose that there are only *two* ingredients in limited supply, and that the range of policies open to the baker can now be described by a feasible region defined by two constraints, one for each ingredient:

$$\begin{aligned} x_1 + 2x_2 &\leqq 4 \\ 2x_1 + x_2 &\leqq 6 \end{aligned} \tag{7–4}$$

As before, x_1 and x_2 (both required to be non-negative) denote the quantities (in dozens) of scones and cakes, respectively, to be made by the baker. Now let us add something to the description of the situation facing the baker: let us suppose that he makes a profit of 2 money units on each dozen scones and (as it happens) exactly the same profit on a dozen cakes, and that these rates of profit are constant (per dozen) no matter how many dozens he makes. Faced with this situation, the baker (we suppose) wants to find the levels of production which make his profit as large as possible. His profit may be expressed (in money units) as a function

$$2x_1+2x_2 \tag{7–5}$$

of the quantities he makes; note that this is a *linear* function because we have supposed that he makes a constant profit of 2 money units on every dozen cakes (and also on every dozen scones). In algebraic terms, his policy problem may now be described as: find non-negative values of x_1 and x_2 which maximize expression (7–5) subject to the constraints (7–4). Any problem which can be formulated in such terms is called a *linear programming problem*; this term indicates that we wish to find non-negative values for a set of variables so as to maximize a *linear* function of the variables subject to *linear* inequalities on the variables. The above problem may be written in matrix notation:

find a vector $\mathbf{x} \geqq \mathbf{0}$ which

maximizes $\begin{bmatrix} 2 & 2 \end{bmatrix} \begin{bmatrix} x_1 \\ x_2 \end{bmatrix}$

subject to $\begin{bmatrix} 1 & 2 \\ 2 & 1 \end{bmatrix} \begin{bmatrix} x_1 \\ x_2 \end{bmatrix} \leqq \begin{bmatrix} 4 \\ 6 \end{bmatrix}$

The general problem may be written:

(7–6)

find a vector $\mathbf{x} \geqq \mathbf{0}$ which maximizes $\mathbf{c}'\mathbf{x}$
subject to $\mathbf{Ax} \leqq \mathbf{b}$
where $\mathbf{c}$, $\mathbf{b}$ and $\mathbf{A}$ are constants.

For the particular constraints of the example, we have already seen (in exercise 2 of section 7.3) that we may define two new variables x_3 and x_4 as the amounts not used up of the two ingredients, and hence that we may write the constraints as

$$\begin{aligned} x_1+2x_2+x_3 \quad &= 4 \\ 2x_1+x_2 \quad +x_4 &= 6 \end{aligned} \tag{7–7}$$

Thus the problem may be written as one of finding non-negative values for x_1, x_2, x_3 and x_4 which maximizes the same function as before, but now subject to a pair of *equality* constraints. (Of course, the four variables are not all independent; for example, once values are chosen for x_1 and x_2, this automatically determines the values of x_3 and x_4.) For the general problem, it is always possible similarly to define new variables to allow the constraints to be written $\mathbf{Ax} = \mathbf{b}$.

When the problem is written with equality constraints, it may be shown that those basic solutions (to the set of constraint equations) which have non-negative values for all the variables correspond to 'corners' of the feasible region. This is a fundamental and most important result in the theory of linear programming. It is not proved here, but the result has already been illustrated for the present example, by exercise 2 of section 7.3. Let us examine this with the aid of Figure 7.1 which shows the feasible region with its four

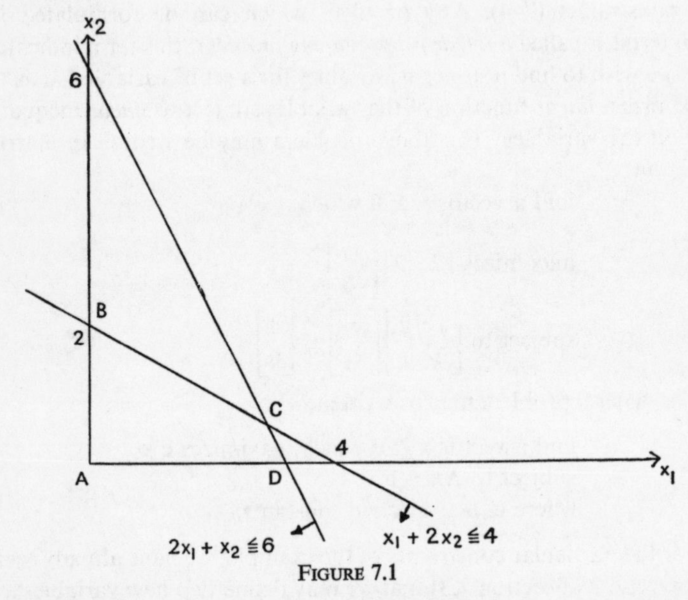

FIGURE 7.1

corners marked A, B, C and D. As seen in the previous exercise, there are six basic solutions to the equations (7–7), and four of these have all elements non-negative. These four are

$[0 \quad 0 \quad 4 \quad 6]$ which corresponds to point A

$[0 \quad 2 \quad 0 \quad 4]$ which corresponds to point B

$[\frac{8}{3} \quad \frac{2}{3} \quad 0 \quad 0]$ which corresponds to point C

$[3 \quad 0 \quad 1 \quad 0]$ which corresponds to point D

Once the feasible region is thus described, the next step is to find the point in the feasible region which corresponds to the production levels x_1 and x_2 which maximize the baker's profits. To explore this, ignore the feasible region for a moment, and consider which alternative values of x_1 and x_2 would together lead to a total profit of (say) 2 money units. Some obvious answers are $(x_1, x_2) = (1, 0)$ or $(0, 1)$; more generally the same profit is given by any point on the line which passes through these two points, i.e. the line

$$2x_1 + 2x_2 = 2$$

(where the left-hand side is simply the profit function itself). We call such a line an *iso-profit line* (since it connects points all corresponding to the same profit level, here 2 money units). Another iso-profit line, connecting all points corresponding to a total profit of 3 money units, is

$$2x_1 + 2x_2 = 3$$

Plot these two lines on the graph of Figure 7.1 and observe that they are parallel to each other, with the latter line further out from the origin than the former. More generally, all the iso-profit lines for this example comprise a family of parallel lines (of slope -1) and the further the line from the origin (in the north-east direction) then the greater is the profit to which the line corresponds.

This suggests an elementary way for finding the point in the feasible region which corresponds to the maximum profit. Consider the family of iso-profit lines, and select the outermost of these (i.e. corresponding to the highest profit) which still intersects the feasible region. Experiment with the graph soon shows that the line through the corner point C is the best that can be done. This is the point

$$(x_1, x_2) = (\tfrac{8}{3}, \tfrac{2}{3})$$

and corresponds to a profit of 20/3 money units. Thus in this example the *optimal solution* (i.e. the values for the $\mathbf{x}$ vector which give the maximum profit) is to be found at a corner point. This result is completely general, although this is not proved here. (There is a small qualification: if the iso-profit lines happen to have the same slope as one of the constraint lines, then the optimal solution may not be unique. For example, if in Figure 7.1, the iso-profit line had the same slope as the first constraint line, which passes through B and C, then these two corner points and all points on the line between them would be equally good. However, in such a case, it still remains true that one or other of the corner points is as good as can be found. Incidentally, if the function to be maximized is not linear, but the constraints are still linear, the optimal solution can no longer by guaranteed to occur at a corner point.)

In summary, we have argued so far that the optimal solution always occurs at a corner point of the feasible region, and that each such corner point (or *extreme point*, as it is usually called) corresponds to a basic solution (for the constraint equations) which has all its elements non-negative. For a small example such as that considered in this section, it is sometimes possible to find the optimal solution by a graphical approach. For larger problems (with more variables) this is inadequate. Instead an algebraic approach is necessary. Rather than search the whole of the feasible region, it is sufficient to examine the *basic feasible solutions* (i.e. the basic solutions with non-negative elements, for the system of constraint equations) since the optimum can be found at one (or perhaps more) of these. Thus only a *finite* number of points need to be examined, and this greatly simplifies the problem. A number of algebraic procedures have been devised for finding the optimal solution to such linear programming problems; almost all of these are ways of *systematically* searching among the basic feasible solutions, so that usually only a very small proportion of them need to be examined before the optimum is reached. Such procedures (or *algorithms*) are highly efficient. When used with electronic computers, they permit the solution of very large problems (e.g. having hundreds or even thousands of variables); and on a few occasions, problems with over one million variables have been tackled successfully. This great computational power has encouraged the widespread use of linear programming models to represent certain problem situations which call for policy decisions. But it is necessary to revert to the warning given at the beginning of the section. In the example of the baker, the use of *linear* constraints seems eminently reasonable; but to represent the baker's aim as the maximization of a *linear* profit function is valid only if (a) the baker really does want to maximize profits, and (b) it is reasonable to suppose that the profit rate (per dozen) is a constant; if it were not, then the profit function would not be linear, and linear programming could not be used. More generally, computational power should not lead us to overlook the fact that the so-called 'optimal solution' is of value to us only if we have created a valid representation of the decision problem, in all senses of the term 'valid representation'; in particular (but by no means exclusively) this requires that we should not use linear programming if the problem cannot be represented (at least to an adequate approximation) by *linear* expressions.

7.5 *Exercises*

*1. A manufacturer makes two products using two inputs, of which his supply is 40 and 20 units respectively. Each unit of the first product requires 2 units of the first input, but does not require any of the second input. Each unit of the second product requires 1 unit of each of the inputs. He makes a profit of one money unit on each unit of the first product, and a profit of 2 money units on each unit of the second product. He wishes to choose production levels so as to maximize his profits. Solve the manufacturer's problem by formulating a linear programming problem, and finding the optimum by examination of the basic feasible solutions. Check your result by preparing and exploring a graph.

*2. Suppose a problem has been formulated as:

choose non-negative values for x_1, x_2 to minimize $2x_1 - x_2$ subject to

$$x_1 + x_2 \leqq 4$$
$$2x_1 + x_2 \geqq 1$$

How can this be converted to the style of formulation given in expression (7–6), where the function is to be *maximized*, and the inequalities are 'less than or equal to'?

7.6 *Quadratic forms*

This section gives a short introduction to a subject which is of importance in certain mathematical and statistical techniques which are applicable in social science contexts. Another reason for introducing quadratic forms is to give an illustration of the use of matrix concepts in handling *non-linear* algebraic expressions; thus this section is an exception to the main purpose (announced in section 1.1) of dealing only with linear algebra.

In scalar algebra, an example of a quadratic term is ax^2 where, as usual, x is a variable and a is a constant coefficient. In this section we are concerned with the generalization of such a term to more than one variable. If we consider just two variables x_1 and x_2, then an example of a quadratic form is

$$ax_1^2+2bx_1x_2+cx_2^2$$

where a, b and c are scalar coefficients. Check for yourself that this expression is the same as

$$[x_1 \quad x_2]\begin{bmatrix} a & b \\ b & c \end{bmatrix}\begin{bmatrix} x_1 \\ x_2 \end{bmatrix}$$

Note that in putting the expression into matrix notation, the coefficients matrix which is used is symmetric. Actually there is some choice in the matter; any matrix can be used provided that (in the case of two variables) the sum of the elements off the principal diagonal is equal to the coefficient of the term involving the product x_1x_2. As an example check that

$$x_1^2+3x_1x_2+4x_2^2$$

can be written out in matrix terms using any one of the following matrices:

$$\begin{bmatrix} 1 & 1 \\ 2 & 4 \end{bmatrix} \quad \begin{bmatrix} 1 & 0 \\ 3 & 4 \end{bmatrix} \quad \begin{bmatrix} 1 & 4 \\ -1 & 4 \end{bmatrix} \quad \begin{bmatrix} 1 & \frac{3}{2} \\ \frac{3}{2} & 4 \end{bmatrix}$$

However, it is conventional and convenient to choose the symmetric matrix. These remarks about the case of two variables generalize to the case of n variables, and lead to the general definition:

Definition. For a vector of n variables $[x_1 \; x_2 \ldots . x_n]$, a *quadratic form* is the expression $\mathbf{x}'\mathbf{A}\mathbf{x}$, where $\mathbf{A}$ is any symmetric matrix of order n.

Note that the expression is a scalar, and that it is homogeneous in the sense that each term involves the variables to a total power of 2; $\mathbf{A}$ is said to be the matrix associated with the form.

Examples:

(1) $x_1^2+2x_2^2+5x_3^2+2x_1x_2+8x_1x_3+4x_2x_1+2x_2x_3$ is a quadratic form in the three variables x_1, x_2 and x_3. Note first that the terms in x_1x_2 and x_2x_1 may be added together (by virtue of the usual

commutative property of *scalar* algebra), and hence the form may be written

$$x_1^2+2x_2^2+5x_3^2+6x_1x_2+8x_1x_3+2x_2x_3$$

The symmetric matrix **A** associated with the form is

$$\begin{bmatrix} 1 & 3 & 4 \\ 3 & 2 & 1 \\ 4 & 1 & 5 \end{bmatrix}$$

(2) $x_1^2+7x_2^2$ is a quadratic form which may be written

$$[x_1 \quad x_2]\begin{bmatrix} 1 & 0 \\ 0 & 7 \end{bmatrix}\begin{bmatrix} x_1 \\ x_2 \end{bmatrix}$$

When a quadratic form arises in practice, we often want to know how the sign of its value behaves as the values of the variables change. For example, the very simple quadratic form $x_1^2+7x_2^2$ is always positive (provided at least one variable is non-zero), but $x_1^2-x_2^2$ is sometimes positive (e.g. if $x_1>x_2>0$) and sometimes negative (e.g. if $x_2>x_1>0$).

Definition. A quadratic form is said to be

- *positive definite* if its value is positive (i.e. >0) for all **x** except $\mathbf{x}=\mathbf{0}$,
- *positive semi-definite* if its value is non-negative (i.e. $\geqq 0$) for all **x** except $\mathbf{x}=\mathbf{0}$,
- *negative definite* if its value is negative (i.e. <0) for all **x** except $\mathbf{x}=\mathbf{0}$,
- *negative semi-definite* if its value is non-positive (i.e. $\leqq 0$) for all **x** except $\mathbf{x}=\mathbf{0}$.

If the form has a positive value for some **x** and a negative value for other **x**, it is said to be *indefinite*.

Examples:

(3) $x_1^2+x_2^2$ is positive definite since its value is positive for all values of x_1 and x_2 except when both are simultaneously zero. (Note that in the definitions, the phrase 'for all **x** except $\mathbf{x}=\mathbf{0}$' does not rule out some x_i being zero; in the present case, if $x_1=0$ and $x_2=-7$, the form has value 49. But when all the variables are zero, *any* quadratic form has zero value; this case is of no interest and that is why it is ruled out in the definition.)

(4) $x_1^2+2x_1x_2+2x_2^2$ is not quite so easy to classify. While in the previous example, each term was a squared expression and hence positive for non-zero $\mathbf{x}$, the cross-product term in the present form is negative if the two variables are of opposite sign; it then becomes a question of comparing that term with the universally positive terms x_1^2 and $2x_2^2$. Fortunately, a very simple piece of manipulation resolves the issue: we may 'complete the square' (as in solving a quadratic equation in one variable) and thus write

$$x_1^2+2x_1x_2+2x_2^2 = x_1^2+2x_1x_2+x_2^2+x_2^2$$
$$= (x_1+x_2)^2+x_2^2$$

Thus we immediately see that the form is positive definite.

More generally, *any* quadratic form may be converted into a sum of squared terms, and hence it may be classified; these results are not proved here, but we now explore an alternative way of approaching the problem, to give some further insight. Given a vector $\mathbf{x}$ of order n, let us transform the variables by defining a new vector of variables $\mathbf{y}$ (also of order n) by the expression

$$\mathbf{y} = \mathbf{Q}\mathbf{x}$$

This is known as a *linear transformation* of the variables. The matrix $\mathbf{Q}$ is square, and we shall deal only with cases where it is non-singular. Thus for given $\mathbf{Q}$, and for any given set of values for the components of $\mathbf{x}$, we get a unique set of values for the components of $\mathbf{y}$. Similarly, since $\mathbf{x} = \mathbf{Q}^{-1}\mathbf{y}$, any given vector $\mathbf{y}$ leads to a unique vector $\mathbf{x}$. (This property identifies what is called a *one-to-one transformation.*) Now return to a given quadratic form $\mathbf{x}'\mathbf{A}\mathbf{x}$. With this transformation of variables, we may write

$$\mathbf{x}'\mathbf{A}\mathbf{x} = (\mathbf{Q}^{-1}\mathbf{y})'\,\mathbf{A}(\mathbf{Q}^{-1}\mathbf{y})$$
$$= \mathbf{y}'(\mathbf{Q}^{-1})'\mathbf{A}\,\mathbf{Q}^{-1}\mathbf{y}$$

Now let $\mathbf{B}$ denote $(\mathbf{Q}^{-1})'\mathbf{A}\mathbf{Q}^{-1}$. Then $\mathbf{B}$ is symmetric (as may be shown by using the same approach as in exercise 12 of section 3.15). Furthermore, we can find a matrix $\mathbf{Q}$ which makes $\mathbf{B}$ a diagonal matrix; this result is established in the following theorem (stated here without proof):

Theorem 7.1. If $\mathbf{A}$ is a symmetric matrix, there exists at least one (square) matrix **R** such that $\mathbf{R'AR} = \mathbf{D}$ where **D** is a diagonal matrix.

Note that **R** may be associated with $\mathbf{Q}^{-1}$ in the previous discussion.

Example:

(5) Given the quadratic form used in example (4),

$$\mathbf{x'Ax} = x_1^2 + 2x_1x_2 + 2x_2^2 = [x_1 \; x_2]\begin{bmatrix} 1 & 1 \\ 1 & 2 \end{bmatrix}\begin{bmatrix} x_1 \\ x_2 \end{bmatrix}$$

consider the transformation $\begin{bmatrix} y_1 \\ y_2 \end{bmatrix} = \begin{bmatrix} 1 & 1 \\ 0 & 1 \end{bmatrix}\begin{bmatrix} x_1 \\ x_2 \end{bmatrix}$

Let this transformation be denoted $\mathbf{y} = \mathbf{Qx}$;

then $$\mathbf{Q}^{-1} = \begin{bmatrix} 1 & -1 \\ 0 & 1 \end{bmatrix}$$

(which confirms that **Q** is non-singular in this case) and

$$(\mathbf{Q}^{-1})'\mathbf{A}\mathbf{Q}^{-1} = \begin{bmatrix} 1 & 0 \\ 0 & 1 \end{bmatrix}$$

a particularly simple diagonal matrix. Thus, with this transformation of variables,

$$\begin{aligned} \mathbf{x'Ax} &= \mathbf{y}'(\mathbf{Q}^{-1})'\mathbf{A}\mathbf{Q}^{-1}\mathbf{y} \\ &= \mathbf{y}'\begin{bmatrix} 1 & 0 \\ 0 & 1 \end{bmatrix}\mathbf{y} \\ &= y_1^2 + y_2^2 \end{aligned}$$

Note that the transformation may be written in scalar terms as

$$y_1 = x_1 + x_2 \text{ and } y_2 = x_2$$

Thus the previous equation is equivalent to writing the form as

$$\mathbf{x'Ax} = y_1^2 + y_2^2 = (x_1 + x_2)^2 + x_2^2$$

In other words, the task of finding the transformation $\mathbf{y} = \mathbf{Qx}$ is the same as that carried out in example (4), namely expressing the quadratic form as the sum of squared terms.

More generally, Theorem 7.1 implies that for any quadratic form, a transformation $\mathbf{y} = \mathbf{Qx}$ can be found to enable us to write $\mathbf{x'Ax} =$

$\mathbf{y}'\mathbf{D}\mathbf{y}$ where $\mathbf{D}$ is some diagonal matrix. We call this 'diagonalizing the quadratic form', or putting it in *diagonal form*. Once we have found $\mathbf{Q}$ and hence found out how to diagonalize the quadratic form, then the form may be classified immediately, as illustrated above in example (4). In small cases, we can usually find $\mathbf{Q}$ by inspection. For more difficult cases, a systematic approach is available; this amounts to carrying out certain numerical operations on the elements of the matrix $\mathbf{A}$; but the details are not discussed here. Note that for a large matrix $\mathbf{A}$, the computation is not trivial – as usual!

7.7 *Exercises*

1. Express the following matrix products as quadratic forms in scalar notation:

   *(a) $[x_1 \quad x_2]\begin{bmatrix} 1 & -2 \\ -2 & 5 \end{bmatrix}\begin{bmatrix} x_1 \\ x_2 \end{bmatrix}$

   (b) $[x_1 \quad x_2]\begin{bmatrix} -1 & 1 \\ 1 & -6 \end{bmatrix}\begin{bmatrix} x_1 \\ x_2 \end{bmatrix}$

2. Express each of the following quadratic forms as a matrix product involving a symmetric coefficients matrix:

   *(a) $x_1^2+2x_1x_2-2x_2^2$
   (b) $x_1^2-2x_1x_2+x_2^2$
   (c) $x_1^2+2x_1x_2+4x_2^2-2x_2x_3+x_3^2$

*3. Diagonalize each of the quadratic forms given in exercises 1 and 2 above, and hence classify each form as positive definite, negative definite, etc.

7.8 *The need for integer-variable analysis*

The mathematical analysis has been developed so far on the assumption that we may employ continuous variables; this applies with especial force to Chapter 6 dealing with the solution of systems of equations. But is has also been remarked that applied contexts sometimes require that the variables take on *integer* values only; indeed such circumstances arise particularly frequently in social science contexts. Accordingly, in the next few sections, we turn to the study of some mathematical analysis involving discrete variables which take on integer values only.

As before, the first remark to be made is that even though we need an answer in integers it may not be necessary to perform its derivation in terms of discrete variables. For example, our baker, who has limited supplies of ingredients and who wishes to plan his production of scones and cakes, may work out his sums in continuous variables and come up with the result that he should make 237·8 scones and 129·2 cakes. There is no reason why he should not use these results. He may round them to 237 scones and 129 cakes, and get on with his baking. The policy he adopts may only be approximately correct (as a result of the rounding) but the error is too small to be a matter of concern, especially relative to any errors of measurement and some uncertainty as to *exactly* how much flour (say) will be required on average to make a scone. (But there *may* still be some difficulty; even though the rounding of the answer has little effect on his profit – supposing that is what he wishes to maximize – it may upset the constraint inequalities; for this reason, the figures have been rounded *down* in the present example, to make sure that the policy does not imply the use of more of the ingredients than is available; in other examples, these difficulties may be more acute.)

In many other contexts, rounding of fractional answers is inadequate. For example, we might wish to schedule certain airline services using a fleet of five identical aircraft, and we might define x_3 as the number of aircraft to be flown from airport A to airport B on each Monday morning. A 'solution' which gives $x_3 = 1{\cdot}48$ may not be at all helpful. To round it to $x_3 = 1$ or $x_3 = 2$ may make a large proportionate difference to the profit (or whatever it is we are interested in) and is likely to upset the constraint inequalities, again especially because the rounding error is a large proportion of the total fleet of aircraft. Another example is the personnel assignment problem (already studied as example (7) in section 3.4). Joe Smith may feel hurt if 0·73 of him is assigned to 0·73 of job 1 and the other 0·27 of him is assigned to job 7. (Remember the underlying assumption that one man can do only one job at once.) And to consider rounding 0·73 to either 1 or 0 means we are back at the beginning of the problem, trying to decide whether or not Joe Smith should do that job. In these sorts of circumstance, we have to recognize at the outset that we need integer answers. Usually this means that we have to try to find some mathematical analysis which will enable us to do this. On occasion however the problem is such that we can work with continuous variables and use mathematical analysis based on continuous variables, secure in the knowledge that all will turn out well in the end, i.e. that even though the mathematics permits fractional answers, those we will actually find at the end of the calculation will all be integers. An example of this happy type of situation is considered later (in section 7.14).

Finally note that the applied context sometimes forces us to formulate what is often called a *mixed-integer* model, in which some of the variables (like the number of scones) may be regarded as continuous variables, at least for purposes of practical calculation, while others (like the number of aircraft) must be regarded as integer variables. In other words, the model has a mixture of continuous and integer variables.

7.9 *The single linear equation in one or two integer variables*

The remarks in the preceding section apply to all models, whether linear or non-linear. The following discussion deals only with linear expressions. Just as in dealing with linear equations in continuous variables (in Chapter 6), we are concerned with whether or not a solution exists; if so, whether it is unique; and, above all, how to calculate such solutions as do exist. In this and the following sections, lower-case italic letters such as a, b etc. will be used to denote integers (which are of course a subset of all scalars). The mathematical theory of equations in integers is not very fully developed, even when attention is confined to linear equations. However, we shall find it useful to look at some of the general theory for small systems before going on to consider computational aspects of large systems involving many variables and many equations.

Let us begin with the simplest system of all – one equation in one integer variable, x:

$$ax = b$$

(Remember that here a and b are themselves supposed to be integers. In practical work this is no restriction; suppose that the coefficients had been measured to two decimal places, to give an equation $12{\cdot}29x = 143{\cdot}47$; this may be multiplied by 100 to yield an equation with integer coefficients.) To establish whether this equation has an integer solution, we must distinguish separate cases (just as in section 6.1):

(i) if $a=0$, then the result depends on whether b is zero:
   (a) if $b=0$, there are an infinite number of solutions; any integer value for x (positive, negative, or zero) makes the left-hand side zero, and the right-hand side is zero,
   (b) if $b \neq 0$, there is no solution, since the left-hand side is always zero while the right-hand side is non-zero,

(ii) if $a \neq 0$, then both sides may be divided by a to give

$$x=b/a$$

Thus the equation has an integer solution if and only if b/a is an integer, i.e. iff a is a factor of b; note that if a solution exists, it is unique. (In practice, we are often interested in non-negative integer solutions; for a solution of this type to exist, there is the additional condition that a and b should have the same sign.) For example, $7x=14$ has a unique (integer) solution $x=2$ because 7 is a factor of 14, i.e. when 7 is divided into 14 the answer is an integer, here 2.

Before going on to consider the next case (of a single equation in two variables), let us revise a concept met in school arithmetic, namely the concept of the *highest common factor* (sometimes known alternatively as the *greatest common divisor*, a more precise term). As an example, consider the numbers 6 and 12, and write down all the *positive* (integer) factors of each number, in ascending order:

6: 1, 2, 3, 6
12: 1, 2, 3, 4, 6, 12

The highest common factor is the largest of these integers which is common to both lists of factors; in this example, it is 6.

Now consider a *homogeneous* linear equation in two integer variables x and y:

$$ax+by=0 \tag{7–8}$$

If the highest common factor (h.c.f.) of a and b is greater than 1, we start by dividing both sides of the equation by the h.c.f. For example, if the equation is

$$4x+8y=0$$

we may divide by the h.c.f. (which is 4) to obtain

$$x+2y=0$$

which is still in the same form as (7–8). Thus, without loss of generality, we may assume that (7–8) is already in a form where a and b have unity as their h.c.f. (We also suppose that both a and b are non-zero; otherwise the equation would not be a two-variable case.) The equation (7–8) may be written

$$x = -\frac{b}{a}y$$

Clearly x is an integer if and only if y takes on a value which is a multiple of a. Every such y is a member of the series

$$y = at$$

where t is any integer (positive, negative or zero). Thus the complete solution is

$$x = -bt \quad \text{and} \quad y = at$$

Note that a solution always exists, and that it is not unique; in fact there are an infinite number of solutions, and $x = 0$, $y = 0$ is always one of them (obtained by setting $t = 0$ in the general solution). If, in some practical context, we are interested in non-negative solutions, then (apart from the 'trivial' solution $x = y = 0$) a non-negative solution exists if and only if a and b are of opposite sign.

Examples:

(1) The equation $x + 2y = 0$ is already in a form where the h.c.f. of its coefficients is unity. The general solution is $x = -2t$, $y = t$. *Part* of the set of solutions may be listed thus:

x:	6	4	2	0	−2	−4
y:	−3	−2	−1	0	1	2

Thus for negative t, y is negative and x is positive; and vice-versa.

(2) The equation $6x - 10y = 0$ has 2 as the h.c.f. of its coefficients. Divide by 2 to yield $3x - 5y = 0$. The general solution is $x = 5t$, $y = 3t$. Note that each positive integer t yields a positive solution.

Now consider the non-homogeneous equation

$$ax + by + c = 0 \tag{7–9}$$

where a, b and c are all non-zero integers. Here we shall see that an integer solution does not always exist, but that when it does, the equation has an infinite number of such solutions, in much the same way as before. These results are now stated formally.

Theorem 7.2. The equation $ax+by+c=0$ (where a, b and c are non-zero integers) has an integer solution for the variables x and y if and only if d (which denotes the highest common factor of a and b) is also a factor of c. If this condition holds, there are an infinite number of solutions, and they are all generated by

$$x = x_0+\frac{b}{d}t, \qquad y = y_0-\frac{a}{d}t$$

where (x_0, y_0) is any one solution, and t is an arbitrary integer.

A complete proof is not offered here. But some parts of the argument can be stated very simply. Let $a=a_1d$ and $b=b_1d$ where a_1 and b_1 must be integers (because d is the h.c.f. of a and b). The equation may be written

$$(a_1x+b_1y)= -c/d$$

Clearly an integer solution exists only if the right-hand side is an integer, i.e. only if d is a factor of c, and this establishes a necessary condition. (Although we have not proved that it is also a sufficient condition, experiment with some numerical cases makes this latter result plausible.) Now suppose that a solution (x_0, y_0) has been found; and let (x, y) denote any other solution. Then

$$a_1x_0+b_1y_0 = -c/d \text{ and } a_1x+b_1y= -c/d$$

Subtract one equation from the other, to give

$$y-y_0 = -\frac{a_1}{b_1}(x-x_0)$$

Now a_1 and b_1 have no common factor (other than unity); this follows because they were obtained by dividing a and b by the *highest* common factor; in other words all the common factors have already been divided out. Now we want $(y-y_0)$ to be an integer, and this will be so only if $(x-x_0)$ is a multiple of b_1. In other words we have $x-x_0=b_1t$ where t is an arbitrary integer. Thus the general solution (x, y) may be written

$$x = x_0+b_1t = x_0+\frac{b}{d}t$$

and

$$y = y_0-\frac{a}{d}t$$

This proves that every solution has this form. We can also show that every pair of values given by this form must be a solution: let (x_1, y_1) be a pair of values given by setting $t = t_1$ in this expression, in other words

$$x_1 = x_0 + \frac{b}{d} t_1$$

$$y_1 = y_0 - \frac{a}{d} t_1$$

Then $$ax_1 + by_1 + c = ax_0 + by_0 + c + \frac{t_1}{d}(ab - ba)$$
$$= ax_0 + by_0 + c$$
$$= 0 \text{ since } (x_0, y_0) \text{ is a solution}$$

Thus (x_1, y_1) must also be a solution.

Note that Theorem 7.2 tells us whether or not the equation has a solution, and also how to generate all solutions once one solution has been discovered; it does not tell us how to find a solution in the first place; there is a systematic method for doing this, but it is not given here. Note also that the general form of solution given in Theorem 7.2 includes the previous result for the homogeneous equation as a special case, because, as we know, the trivial solution $x = y = 0$ always exists for the homogeneous case, and if $x_0 = y_0 = 0$ is substituted in the general form given in Theorem 7.2, then the general solution becomes

$$x = (b/d)t \quad \text{and} \quad y = -(a/d)t$$

which is effectively the same as before; the earlier expressions supposed that the h.c.f. had already been divided out, i.e. that $d = 1$.

Examples:

(3) For the equation $6x + 12y = 24$, the h.c.f. of the coefficients on the left-hand side is 6, and this is a factor of 24. Thus the equation has integer solutions. By inspection, we note that the coefficient of y is itself a factor of 24; hence one solution is $x = 0$, $y = 2$. Then the general solution is

$$x = 2t, \quad y = 2 - t$$

where t is an arbitrary integer.

(4) For the equation $6x+3y=7$, the h.c.f. is 3; this is not a factor of 7. Thus (by Theorem 7.2) there is no solution.

(5) For the equation $3x-5y=7$, the h.c.f. is 1, and this is a factor of 7. Thus (by Theorem 7.2) solutions exist and are given by $x=x_0-5t$, $y=y_0-3t$. To find one solution, write $x=(5y+7)/3$ and consider series of possible values $y=0, 1, 2, \ldots$ until one is found which gives an integer value for x. The value $y=1$ gives $x=4$; in other words this is one solution, and the general solution is

$$x=1-5t, \quad y=4-3t$$

(Note that all non-positive values for t give a positive solution for both x and y.)

7.10 *Exercises*

1. For each of the integer-variable equations $ax=b$, $ax+by=0$ and $ax+by+c=0$, compare the results given in the previous section with those obtained in Chapter 6 for the corresponding cases involving continuous variables. Consider graphical representations of the various cases. What general remark can be made to compare the number of solutions in an integer variable case with the number for the corresponding case in continuous variables?

*2. If x and y are integer variables, find all solutions (if any) for each of the following equations:

(a) $2x-y=0$
(b) $x+y=3$
(c) $3x+9y=13$

*3. Consider the system of two *simultaneous* linear equations in integer variables, comprising the first two equations of exercise 2. Does this system have a solution? Also consider the system graphically and compare it with the system of the same two equations involving continuous variables.

*4. Does the following system have an integer solution?

$$x-3y=0$$
$$2x+y=5$$

7.11 *Larger systems of linear equations in integers*

This section is devoted to systems of *simultaneous* linear equations in integer variables. In general, the chances of such a system having a solution or solutions are rather less than if the same set of equations involved continuous variables. Graphically, the solutions (if any) to a single linear equation in two integer variables comprise a set of equally spaced points lying along a line; in the corresponding continuous variable case, the set of solutions is *all* points on the line. A system of two such equations has a unique integer solution if and only if the two lines intersect *and* the point of intersection has integer coordinates (as in exercise 3 of the previous section); where the lines intersect, the corresponding case in continuous variables always has a solution. To give further illustration, and to show one way of attempting to solve a system of linear equations in integers, let us now consider a larger-scale example.

Example:

(1) Consider the following system of equations which (we suppose) represents a situation in which all variables must be strictly positive integers (i.e. > 0):

$$\begin{aligned} x_1+x_2+x_3+x_4 &= 12 \\ -x_1 \qquad +x_3 \qquad &= \ \ 7 \\ x_1+x_2 \qquad\qquad &= \ \ 2 \end{aligned}$$

For the moment let us ignore the positive integer requirement, and simply regard the system as one in continuous variables (which may be positive or negative). If elementary row operations are applied in the usual way to the augmented matrix, we find that $r(\mathbf{A}) = r(\mathbf{U}) = 3$, the equations belong to Case III of Table 6.11, one variable may be assigned a parametric value, and the general solution is (for any value of θ)

$$[x_1 \quad x_2 \quad x_3 \quad x_4] = [(3-\theta) \quad (\theta-1) \quad (10-\theta) \quad \theta]$$

Now suppose we recognize the requirement that each variable must be strictly positive. This still leaves an infinite number of solutions, since the previous expression holds for all θ in the range $1 < \theta < 3$. Finally, if we recognize the integer requirement as well, there is only one solution, given by $\theta = 2$, which is

$$[x_1 \quad x_2 \quad x_3 \quad x_4] = [1 \quad 1 \quad 8 \quad 2]$$

Of course, it is possible to spot this result by inspecting the equations. If $x_1+x_2=2$, then the only positive integers which satisfy this equation are $x_1=x_2=1$, and this immediately leads to the unique solution already obtained. (This kind of observation sometimes works even with non-linear equations in integers; see exercise 2 in the next section, which is the kind of problem often set in newspapers and magazines as 'brain-teasers'; exercise 3 gives an applied context for another such problem, though this time it is linear.)

Although the integer requirement generally reduces the number of solutions compared with the corresponding case for continuous variables, it does not (of course) necessarily lead to a unique solution or to no solution at all. This may be illustrated quickly: if in the previous example, the variables are required to be *non-negative* integers (i.e. now including zero), then the parameter θ can take on any integer value in the range $1 \leqq \theta \leqq 3$; in other words θ can be 1, 2 or 3 and there are three distinct solutions. In a case such as this, where the integer requirement leads to only a small number of solutions, the task of finding the solutions may be made relatively easy; in particular, it may be computationally feasible just to list all potential solutions and to examine each in turn to find out which are in fact solutions. This approach by *enumeration* is developed in section 7.13; see also exercise 1 in the next section.

7.12 *Exercises*

1. For the system

$$\begin{aligned} x_1+4x_2+x_3 &= 5 \\ x_1+\ x_2 \qquad &= 2 \end{aligned}$$

in non-negative integer variables, find all solutions by alternative methods:

*(a) by first treating the system as one in continuous variables, and then adapting the general solution.

(b) by enumerating all non-negative integer solutions to the second equation and eliminating from this list those solutions which do not also satisfy the first equation.

*2. If all variables are strictly positive integers, find all solutions (if any) for the following non-linear system:

$$\begin{aligned} x_1x_4+x_2x_3 &= 8 \\ x_1x_3+ \quad x_5 &= 2 \\ (x_3+x_5)x_2- \quad x_4 &= 1 \end{aligned}$$

*3. A committee of nine members had to elect a chairman. Each member could vote for any of the three candidates, Andrews, Brown and Clark, and had to state both a first and second choice. In the event, there were no abstentions, and none of the six possible permutations of vote was used more than twice. At the first count (of first choices) Brown had one more vote than Andrews. The rules of the election required that if one candidate got less votes at the first count than *both* the others, he should be ruled out at that stage, and his votes transferred to the other candidates according to the *second* choices marked on the relevant ballot papers. This rule was brought into use, and a chairman elected after the second count. Incidentally, the *total* votes (i.e. first and *all* second choice together) cast for Andrews numbered 7 and those for Clark numbered 6. Who was elected? And how many votes did he get (a) at the first count, and (b) at the second count?

4. Find all non-negative integer solutions (if any) for the following system:

$$\begin{aligned} x_1+\ x_2+2x_3+3x_4 &= 3 \\ x_1-2x_2 \qquad\quad +\ x_4 &= 1 \\ x_1+\ x_2+\ x_3 \qquad\quad &= 2 \end{aligned}$$

7.13 *General integer systems and computational approaches*

The previous sections on the analysis of equations in integer variables may now be reviewed briefly. We have seen that there is a fully developed mathematical theory for a simple case such as that of a single linear equation in two variables. A small system of *simultaneous* linear equations may be tackled by first pretending that it involves continuous variables. If there is a unique solution (on that basis), we can then check whether or not all the values are integers. If there are an infinite number of solutions (in the case of continuous variables) we can then enumerate alternative parameter values and see which (if any) of the list satisfies the integer requirements; example (1) of section 7.11 illustrates this case, as do exercises 1, 3 and 4 of the last section.

In practical contexts, it is systems of *simultaneous* equations which are of interest, and those which occur are not necessarily very small. For such systems, there is not a great deal in the way of developed mathematical theories, and certainly little in the way of elegant, well-defined and computationally simple methods of solution (in contrast to the case of a single linear equation in two integer variables). Thus, in general, it is necessary to fall back on some enumerative approach. Because the variables are in integers, and because in practical circumstances there is both an upper and lower bound on the value of each variable, only a finite number of cases have to be examined in order to test which of them constitute solutions. For example, if a problem involves four variables, each of which is constrained to be a non-negative integer not greater than 5, then there are 6 possible values (0, 1, 2, 3, 4 and 5) to be considered for each variable, and hence an approach by *complete* enumeration would require examination of 'only' $6 \times 6 \times 6 \times 6 = 6^4 = 1296$ cases; here, each case is a different vector (e.g. $[0 \quad 0 \quad 0 \quad 1]$, $[0 \quad 0 \quad 0 \quad 2]$ and so forth) for the values of the four variables. As this example suggests, complete enumeration requires the examination of a very large number of cases for all but the smallest of examples. However, the development of electronic computers has made such enumeration faster and cheaper to execute. Despite this, complete enumeration is prohibitively expensive (or even impossible) under many practical circumstances. Fortunately, in most cases it is possible to devise a computational scheme of *partial enumeration*, in which only a subset of the total number of cases is examined, the method showing that examination of the remaining cases is unnecessary. The approach which uses elementary row operations on the augmented matrix can be regarded as coming into this category; where the equations (thought of as being in continuous variables) have an infinite number of solutions, it is sometimes the case that only a small set of discrete values for the arbitrary parameters need to be examined to see if they furnish integer solutions; in example (1) in section 7.11, only one such parametric value had to be examined.

In many social science contexts involving non-negative integer variables, the problem is not so much one of solving a set of linear equations but rather one of finding which point or points in a feasible region (determined by linear *inequalities*) maximizes (or minimizes) some function of the variables. If the function to be maximized is itself linear in the (integer) variables, then the problem is exactly the same as the linear programming problem defined by expressions (7–6) in section 7.4, except for the additional requirement that the solution be in integers. With this requirement included, we speak of a linear *integer programming* problem; again partial enumeration procedures may be helpful.

Within this class of linear integer programming problems, there is an important subclass of problems of a kind which occur frequently in social science contexts. These are problems in which each variable takes on the value zero or one; such 'zero–one' variables (as they are called) have already been encountered in the personnel assignment problem described in example (7) of section 3.4. The following is a small example of a linear integer programming problem in zero–one variables:

$$\text{maximize} \qquad x_1+2x_2+x_3-x_4+x_5+x_6$$

$$\begin{aligned} \text{subject to} \qquad & 2x_2+x_3-x_4 \leqq 5 \\ & x_1+x_2+x_3+x_4+x_5+x_6 \leqq 2 \\ \text{all} \quad & x_1 = 0 \text{ or } 1 \end{aligned}$$

Since there are six variables and each can take on either of two values, a complete enumeration would require examination of $2\times2\times2\times2\times2\times2=2^6=64$ cases. However a very simple device can be used to establish a satisfactory procedure of partial enumeration: because of the second constraint, not more than two of the variables can be non-zero, since otherwise the left-hand side of that constraint would exceed the value 2. There are 15 ways of choosing two variables to be non-zero, 6 ways of arranging for just one variable to be non-zero, and one case in which all variables are zero. Thus it is sufficient to examine just these 22 cases, rather than the full list of 64 cases. For integer programming problems in general, partial enumeration procedures are sometimes a useful line of attack. There are still limits to the size of problem which can be tackled in this way even with the aid of electronic computers, but there are now a variety of such procedures (some considerably more subtle than the device just illustrated) which sometimes are good enough to give an economical solution even of quite large problems. The field is still one of active research.

7.14 *Problems where the integer requirement is automatically satisfied*
Towards the end of section 7.8, it was remarked that calculations can sometimes go ahead on the assumption that continuous variables are being used, while an integer solution can nevertheless be guaranteed because of the form of the problem. This is of considerable value in some integer programming problems, as will be illustrated below. However let us begin with an example involving the solution of simultaneous equations:

$$\begin{aligned} x_1 + x_2 + x_3 &= 6 \\ x_1 + 2x_2 + 3x_3 &= 11 \\ x_2 + 3x_3 &= 6 \end{aligned}$$

where the variables are required to be integers. It is in fact possible to determine that this system has a unique solution which happens to be entirely in integers, and this can be settled before the equations are solved. In practice, the amount of arithmetic required to establish the result is almost the same as that required to find a solution. Therefore the method is not given here; instead let us work in terms of continuous variables, and apply elementary row operations in the usual way to the augmented matrix. This shows the system to belong to Case II of Table 6.11 with a unique solution $[x_1\ x_2\ x_3] = [2\quad 3\quad 1]$. In this case we are lucky enough to get an answer which also satisfies the integer requirement. Previous examples (e.g. in section 7.11) show that if the system (regarded as being in continuous variables) has an infinite number of solutions, then at least some of these will not satisfy the integer requirement (because the parameter or parameters in which the general solution is expressed are themselves continuous variables). However in the example solved above, all solutions (i.e. the one and only solution) to the system in continuous variables also satisfy the integer requirement.

As indicated already, this kind of property is particularly important for some integer programming problems. To illustrate this, consider again the personnel assignment problem introduced in section 3.4. Suppose that there are just two jobs and two people; thus we define four zero–one variables, x_{11}, x_{12}, x_{21} and x_{22}. Then we have four equality constraints:

$$\begin{aligned} x_{11} \qquad + x_{21} \qquad &= 1 \\ x_{12} \qquad + x_{22} &= 1 \\ x_{11} + x_{12} \qquad\qquad &= 1 \\ x_{21} + x_{22} &= 1 \end{aligned}$$

Now let us add to the specification of the problem by supposing that if the ith person is assigned to the jth job, he earns p_{ij} money units, and that we wish to assign the two people in such a way as to maximize the sum of their earnings. Algebraically, we want to choose values for the x_{ij} to

maximize $\qquad p_{11}x_{11}+p_{12}x_{12}+p_{21}x_{21}+p_{22}x_{22}$

(Check this for yourself.) We have now formulated a linear integer programming problem in which each x_{ij} must be not merely *any* integer but either zero or one. Since we have not given numerical values to the p_{ij}, we cannot calculate the optimal solution. But we can now prove that it will always turn out to satisfy the integer (or rather the zero–one) requirements even if we treat the problem as an ordinary linear programming problem in continuous variables. To show this, consider the four constraint equations, form the augmented matrix, and apply elementary row operations in the usual way. This shows that the rank is 3, there are an infinite number of solutions (if the variables are regarded as continuous variables) and the general solution may be written

$$[x_1 \; x_2 \; x_3 \; x_4]=[\theta \; (1-\theta) \; (1-\theta) \; \theta]$$

after x_4 has been set equal to a parameter θ. By setting $\theta=0$, we obtain one basic solution, $[0 \; 1 \; 1 \; 0]$. Note that this is degenerate. Further exploration along the usual lines shows that there are in all four basic solutions; all of them are degenerate and it turns out that the four therefore become only two *distinct* solutions, the second being $[1 \; 0 \; 0 \; 1]$. (In practical terms, these two solutions correspond respectively to (a) the first person getting the second job and vice-versa, and (b) the first person getting the first job, and the second person the second job.) Thus all the basic solutions have integer values, and all are feasible (i.e. non-negative). Now if the problem is solved as a linear programming problem in continuous variables, the optimal solution will correspond to a feasible basic solution for these equations (by the argument given in section 7.4), and hence the optimal solution must be entirely in integers. For this problem and for similar linear integer programming problems (in any integers, not just zero–one variables), it is often possible to learn in advance that the solution must be in integers, i.e. that all the basic solutions are in integers. If the constraint equations are denoted $\mathbf{Ax}=\mathbf{b}$, this may often (but not always) be done simply by inspecting the elements in the matrix **A**. Fortunately, many practical problems come into the category where this test *can* be applied. Although the test is a simple one, it is not given here.

7.15 *Exercises*

*1. A car dealer has stocks of 5 and 7 identical sports cars at two wholesale depots. He wishes to move the cars to two garages so as to have 3 cars at the first garage and 9 at the second. The cost of moving a car from the ith depot to the jth garage is c_{ij}, and the dealer wants to know how many cars to move from each depot to each garage so as to meet the requirements at minimum cost. Formulate the problem as one of linear integer programming, and by inspecting feasible basic solutions, show that it may always be solved by treating it as a linear programming problem in continuous variables.

*2. If, in the previous exercise, the costs are $c_{11} = 2$, $c_{12} = 5$, $c_{21} = 4$ and $c_{22} = 6$, find the optimal solution. Examine your result in practical terms. Are you surprised?

APPENDIX A

Determinants

Many older text-books, which give an elementary treatment of linear algebra, introduce determinants at an early stage, and then use them in the development of such concepts as the rank of a matrix and the inverse of a non-singular matrix. More recently, the disadvantages of such an approach have received greater recognition, especially in applied work where the emphasis is on computational efficiency. Although determinants do provide *a* method for computing the inverse and for solving systems of linear equations, such an approach is not very efficient in computational terms, and (in particular) it is less efficient than the approach employing elementary row operations which was used in Chapters 4, 5 and 6 of this text. Furthermore this latter approach permits the derivation of constructive proofs for many of the theorems in Chapters 4 and 5, and this has the advantage of integrating the establishment of general properties with the computations used for the application of these concepts in particular numerical cases. However, the student who has mastered the contents of this text-book and who wishes to go on to read articles or books which do employ determinants, will need to know the rudiments of the subject. This appendix is designed accordingly.

A determinant may be defined in a number of alternative ways. Since the concept is a clumsy one, each of these definitions is somewhat complex. We give here a fundamental definition, which is the one which is usually used. But first it is necessary to study some preliminary concepts.

Definition. Any arbitrary ordering of the positive integers 1, 2, 3, ..., k is called a *permutation* of these integers.

In contrast to the concept of a set, the order in which the integers are written down is the crucial point of the definition. Examples of such permutations are:

(1) for the case $k = 4$, the ordering 1, 2, 4, 3
(2) again for the case $k = 4$, the ordering 4, 3, 1, 2
(3) for the case $k = 6$, the ordering 6, 5, 1, 4, 3, 2

Definition. For any selected integer in an ordering or permutation, the number of integers which follow it in the ordering but which precede it in the natural ordering 1, 2, 3, ..., k is called the number of *inversions* caused by the selected integer. If this number is calculated for each integer in the ordering and if these numbers are added together, the result is called the *total number of inversions* which occur in the ordering. If this total is odd the ordering is called an *odd permutation*, and if it is even the ordering is called an *even permutation*; the *signs* of the permutations are said to be negative and positive respectively.

In example (1) above, the number of inversions caused by each integer may be listed:

selected integer	number of inversions
1	0
2	0
4	1
3	0

In this case, the only integer which is followed by an integer out of natural order is the integer 4 which is thus followed by only one integer, namely 3. Thus the total number of inversions is 1 and this is an example of an odd permutation. Check for yourself that example (2) has a total of 5 inversions and hence is an odd permutation, and that example (3) has 12 inversions and hence is an even permutation. The next task is to establish a property relating to permutations.

Theorem A.1. Interchange of two integers in a given permutation changes the sign of the permutation, i.e. if the permutation is odd, it is changed to even, and vice-versa.

Proof. Suppose that the integers being interchanged are in the ith and j^{th} positions (where $j > 1$), and let m denote the number of integers *between* these two positions. The interchange can be regarded as being effected in two stages: first, the jth integer is brought back into a (new) position just before the ith integer, i.e. it passes over $m+1$ integers comprising the integer in the ith position and the m integers between the ith and j^{th} position; in the second stage, the integer in what was the ith position is moved on to the jth position, i.e. it passes over the m integers which are between the ith and j^{th} positions. Now on any occasion when one of the integers is moved over another integer, an inversion is either introduced or removed (see example below). Thus there are a total of $2m+1$ inversion changes, and since this is an odd number, the sign of the permutation changes.

Example:

(4) Suppose the given ordering is 1, 2, 4, 3 (as in example (1)), and suppose that the integers 2 and 3 are interchanged. First 3 is brought back into second position, passing over 4 (which removes an inversion) and over 2 (which creates an inversion). Then 2 is moved on into the fourth position, passing over 4 (which creates an inversion). The original permutation is odd; the new one 1, 3, 4, 2 is even.

Having cleared the decks, we are now ready to introduce the concept of a determinant. Consider any *square* matrix **A**. Associated with this matrix is a *scalar* quantity called a determinant, which is usually denoted $|\mathbf{A}|$. If the elements of **A** are denoted in the usual way, then $|\mathbf{A}|$ may be written out in a fuller notation:

$$|\mathbf{A}| = \begin{vmatrix} a_{11} & a_{12} & \cdots & a_{1n} \\ a_{21} & & & \cdot \\ \cdot & & & \cdot \\ \cdot & & & \cdot \\ \cdot & & & \cdot \\ \cdot & & & \cdot \\ a_{n1} & \cdots & \cdots & a_{nn} \end{vmatrix}$$

As this notation implies, $|\mathbf{A}|$ will be defined in terms of the elements of **A**. After giving a formal definition, we consider its meaning and some implications, by looking at examples.

Definition. The *determinant* $|\mathbf{A}|$ corresponding to a matrix $\mathbf{A} = [a_{ij}]$ of order n is the (scalar) sum

$$\Sigma(\pm) a_{1i} a_{2j} \ldots a_{ns}$$

Each term comprises a product of n of the elements of **A**, and there is one such term for each permutation of the second subscripts. A term is assigned a plus sign in the summation if $i, j, \ldots, s$ is an even permutation of the integers $1, 2, 3, \ldots, n$, and a minus sign if it is an odd permutation.

Example:
(5) Consider

$$\mathbf{A} = \begin{bmatrix} a_{11} & a_{12} \\ a_{21} & a_{22} \end{bmatrix}$$

Then $|\mathbf{A}|$ is defined as the sum of certain products of two elements from **A**. The set of integers $\{1, 2\}$ yields two permutations 1, 2 and 2, 1, which are (respectively) even and odd. Thus there are two terms in the sum

$+a_{11}\, a_{22}$ (corresponding to the first of these permutations) and

$-a_{12}\, a_{21}$ (corresponding to the second). Thus

$$|\mathbf{A}| = +a_{11}\, a_{22} - a_{12}\, a_{21}$$

Remember that although a determinant corresponds to a matrix, it is itself a scalar. Notice that any one term in the summation can be thought of in the following way: choose one element from each row of **A**, and do this in such a way that each column provides exactly one element. (Note also that the unit entries in a permutation matrix – defined in example (7) of section 3.4 – have the same property.) Clearly, for a matrix of order n, there are n ways of choosing an element from the first row, $(n-1)$ ways of choosing an element from the second row (since one column is debarred after an element is chosen from the first row), and so forth. Thus there are $n(n-1)(n-2) \ldots 3.2.1$ different ways of defining a term for the summation, and hence this number measures the number of terms in the summation. The number is known as n factorial, often denoted $n!$

Examples:
(6) If

$$\mathbf{A} = \begin{bmatrix} a_{11} & a_{12} & a_{13} \\ a_{21} & a_{22} & a_{23} \\ a_{31} & a_{32} & a_{33} \end{bmatrix}$$

then

$$|\mathbf{A}| = a_{11}a_{22}a_{33} - a_{11}a_{23}a_{32} + a_{12}a_{23}a_{31} - a_{12}a_{21}a_{33} + a_{13}a_{21}a_{32} - a_{13}a_{22}a_{31}$$

Again note the sign of each term, according to whether the permutation of the second subscripts is even or odd. And note that there are $3! = 6$ terms in the summation.

(7) If

$$\mathbf{A} = \begin{bmatrix} 1 & 3 & -2 \\ 0 & 1 & 4 \\ 5 & 6 & 8 \end{bmatrix}, \text{ then } |\mathbf{A}| = 8-24+60+10 = 54$$

This definition of a determinant does *not* provide an efficient way of computing the value of a determinant. Since there are $n!$ terms, each comprising the product of n elements of $\mathbf{A}$ (when that matrix is of order n), then the amount of arithmetic increases very markedly as n increases; for $n = 4$, there are 24 terms, and for $n = 5$, there are 120 terms. However, this form of the definition does permit certain properties to be established readily. (A more efficient computational procedure is outlined later.)

Theorem A.2. If $\mathbf{B}$ is derived from a square matrix $\mathbf{A}$ by interchanging two rows, then $|\mathbf{B}| = -|\mathbf{A}|$.

Proof. Disregarding the signs for a moment, each term in the summation comprising $|\mathbf{A}|$ is the same as a corresponding term in the summation for $|\mathbf{B}|$; this follows because we make *all possible* choices of one entry from each row and column when compiling the list of terms which go into the summation. In other words, a pair of corresponding terms each has the same n factors; but they are written out in a different order. In particular, one pair of second subscripts is interchanged. Hence (by Theorem A.1), the sign of the permutation is changed, and thus the term is given the opposite sign when it is included in the summation. This applies to each of the $n!$ pairs of terms. Hence $|\mathbf{B}| = -|\mathbf{A}|$.

Example:

(8) Consider

$$\mathbf{A} = \begin{bmatrix} a_{11} & a_{12} \\ a_{21} & a_{22} \end{bmatrix}$$

and let the new matrix obtained by interchanging rows be denoted

$$\mathbf{B} = \begin{bmatrix} b_{11} & b_{12} \\ b_{21} & b_{22} \end{bmatrix} = \begin{bmatrix} a_{21} & a_{22} \\ a_{11} & a_{12} \end{bmatrix}$$

Using the definition of a determinant, the two terms in each of the summations are:

for $\|\mathbf{A}\|$	for $\|\mathbf{B}\|$
$+a_{11}a_{22}$	$-b_{12}b_{21} = -a_{22}a_{11}$
$-a_{12}a_{21}$	$+b_{11}b_{22} = +a_{21}a_{12}$

Thus $|\mathbf{B}| = -|\mathbf{A}|$. (Note that when the terms for $|\mathbf{B}|$ are expressed in the a_{ij} the 'natural' order for the *first* subscripts is 2, 1 as is shown by the expression for **B**.)

Theorem A.3. If **A** is any square matrix, $|\mathbf{A}| = |\mathbf{A}'|$ where (as usual) $\mathbf{A}'$ is the transpose of **A**.

This may be proved in much the same way as the previous theorem. Try out an example for yourself in order to see how to do it. The following theorems are stated and proved, but not illustrated; in each case you should check an example for yourself.

Theorem A.4. If **A** is any square matrix and if **B** is obtained from **A** by multiplying the ith row (*or* column) by a scalar λ, then $|\mathbf{B}| = \lambda|\mathbf{A}|$.

Proof. Consider the case where the ith *row* is multiplied. Each term in the summation comprising $|\mathbf{A}|$ contains exactly one factor from this row. Thus the corresponding term in the summation for $|\mathbf{B}|$ is λ times the term in the summation for $|\mathbf{A}|$. Hence $|\mathbf{B}| = \lambda|\mathbf{A}|$. (When the elements of a column are multiplied, the proof is precisely similar.)

Theorem A.5. If **B** is derived from a square matrix **A** by interchanging two columns, then $|\mathbf{B}| = -|\mathbf{A}|$.

Proof. $\mathbf{B}'$ is derived from $\mathbf{A}'$ by interchanging two *rows*. Hence from Theorem A.2, $|\mathbf{B}'| = -|\mathbf{A}'|$. Now from Theorem A.3, $|\mathbf{B}| = |\mathbf{B}'| = -|\mathbf{A}'| = -|\mathbf{A}|$.

Theorem A.6. If **A** is a square matrix with two identical rows (or columns), then $|\mathbf{A}| = 0$.

Proof. Consider the case where two *rows* are identical. Form a new matrix **B** by interchanging these two rows. Then (by Theorem A.2), $|\mathbf{B}| = -|\mathbf{A}|$. But $\mathbf{B} = \mathbf{A}$, of course. Thus $|\mathbf{B}| = |\mathbf{A}|$. Hence $|\mathbf{A}| = -|\mathbf{A}|$. This can be true only if $|\mathbf{A}| = 0$. (The proof for identical columns is precisely similar.)

Theorem A.7. If **A** is any square matrix and if **B** is obtained from it by replacing row i of **A** with row i itself plus λ times row k, then $|\mathbf{B}| = |\mathbf{A}|$. (A similar result holds for column operations.)

Proof. Consider a typical term from the summation which comprises $|\mathbf{A}|$. Let the factor from row i be denoted a_{ij}. In other words, we suppose that this factor comes from the jth column of **A**. Then in **A** the element in row k and in the same (j^{th}) column is a_{mj}. The corresponding term in the summation comprising $|\mathbf{B}|$ has, instead of the factor a_{ij}, a factor $(a_{ij}+\lambda a_{mj})$. Similar remarks apply to each other pair of corresponding terms in $|\mathbf{A}|$ and $|\mathbf{B}|$. Thus $|\mathbf{B}|$ can be separated into two parts, which equal (respectively) $|\mathbf{A}|$ and $\lambda|\mathbf{C}|$ where **C** is obtained from **A** by replacing row i by row k. Now **C** has two identical rows. Hence by Theorem A.6, $|\mathbf{C}| = 0$. Thus $|\mathbf{B}| = |\mathbf{A}|$.

Theorem A.8. If **A** is a square matrix having one row equal to a non-zero scalar multiple λ of another row, then $|\mathbf{A}| = 0$. (The same result holds for columns.)

Proof. Form a new matrix **B** from the matrix **A** by dividing by λ each element in the row which is λ times the other row. Then $|\mathbf{A}| = \lambda|\mathbf{B}|$, by Theorem A.4. Now **B** has two identical rows, and hence $|\mathbf{B}| = 0$, by Theorem A.6. Thus $|\mathbf{A}| = 0$.

The next theorem deals with the determinant of a triangular matrix (cf. section 3.4). It gives us a way of evaluating determinants, and will also be used later to associate determinants with the concepts of linear dependence and rank.

Theorem A.9. If **B** is an (upper or lower) triangular matrix, then $|\mathbf{B}|$ is the product of all the elements b_{ii} on the principal diagonal.

Proof. Suppose that **B** is *upper* triangular, and of order n. (The proof for the lower triangular case is precisely similar.) Consider the definition of a determinant; one term in the summation of products is

$$+b_{11}\, b_{22} \ldots b_{nn}$$

Each other term must include at least one element from below the principal diagonal. But all such elements are zero, and hence all these product terms are zero. Thus

$$|\mathbf{B}| = \prod_{i=1}^{n} b_{ii} = b_{11} b_{22} b_{33} \ldots . b_{nn}$$

(Here Π means 'multiply', just as Σ means 'add'.)

Now let us consider how this may be used to give a way of evaluating a determinant $|\mathbf{A}|$, where $\mathbf{A}$ is any square matrix. We already know from Chapter 4 that $\mathbf{A}$ may be reduced to an echelon matrix by applying elementary row operations of all three types (as defined at the beginning of section 4.2). Clearly if we use elementary row operations of types (i) and (iii) only, $\mathbf{A}$ may be transformed to an upper triangular matrix; the only difference between this and the corresponding echelon matrix is that the leading unit entries of the echelon matrix will (in general) be values other than unity in the triangular matrix, because we have not used the second type of elementary row operation. The following example illustrates: for the matrix $\mathbf{A}$ shown in Tableau 0 of Table A.1, twice the first row is subtracted from the second row, and three times the first row is subtracted from the third row, the result being shown in Tableau 1; finally, the second and third rows are interchanged to give the upper triangular matrix in Tableau 2. Now the third type of elementary row

TABLE A.1

Tableau 0			Tableau 1			Tableau 2		
1	2	3	1	2	3	1	2	3
2	4	1	0	0	−5	0	2	3
3	8	12	0	2	3	0	0	−5

operation changes the matrix to which it is applied, but the new matrix has the same determinant as the old, by Theorem A.7. Furthermore, the first type of elementary row operation (interchange of rows) does no more than change the sign of the corresponding determinant. Thus we have a simple relationship between the determinant of the original matrix $\mathbf{A}$ and that of the triangular matrix $\mathbf{B}$, namely $|\mathbf{A}| = (-1)^r |\mathbf{B}|$ where r is the number of times we have interchanged a pair of rows. Also, from Theorem A.9,

$$|\mathbf{B}| = \prod_{i=1}^{n} b_{ii}$$

which may be computed readily. In the example given in Table A.1, $|\mathbf{B}| = -10$, there is one interchange of rows in the course of the transformation (i.e. $r = 1$) and hence $|\mathbf{A}| = 10$. (Check this result by applying the definition of a determinant to the original matrix.) This method is a fairly efficient way of evaluating a determinant. Also the argument underlying the method gives a useful way of relating determinants to the concepts of linear dependence and rank. In preparation for this, note that after the second type of elementary row operation has been used to transform the upper triangular matrix into an echelon matrix, then (as we can see from Theorem A.4), the determinant of the resulting echelon matrix is a non-zero multiple of that of the triangular matrix, and hence of that of the original matrix; in other words, if **E** denotes the echelon matrix, $|\mathbf{E}| = \lambda|\mathbf{A}|$ where λ is a non-zero scalar which may be positive or negative.

Theorem A.10. If **G** is a matrix of order $m \times n$, then the rows of **G** are linearly independent if and only if **G** contains a square submatrix **A** of order m whose determinant is non-zero.

Proof. We have already seen (from Chapter 4) that the rows of a matrix are linearly independent if and only if an echelon matrix derived from it has no row comprising zero entries (cf. in particular the discussion in sections 4.6 and 4.10). If a square echelon matrix has no row of zero entries, then (by Theorem A.9) its determinant is unity; whereas if it does have one or more rows of zero entries, its value is zero. The m columns in **G** which correspond to such a square echelon matrix give a determinant of value $\lambda(\neq 0)$ and 0, respectively, according to whether the echelon does not, or does, have one or more rows of zero entries; this follows from the argument preceding this theorem. In other words, a non-zero determinant associated with a set of m columns in **G** corresponds to an echelon matrix having no row comprising zero entries only, and hence the rows of **G** are linearly independent if and only if **G** contains a square submatrix of order m whose determinant is non-zero.

The following theorem is an immediate consequence, because the rank of a matrix is defined as the largest number of linearly independent rows in the matrix:

Theorem A.11. If **A** is a matrix of order $m \times n$, and if the largest square submatrix which has a non-zero determinant, is of order p, then the rank of **A** is p.

In particular, this means that a square matrix **A** is non-singular if and only if its determinant is non-zero. Finally, we give one further property of determinants, a property which is sometimes useful in theoretical analysis. Given a square matrix **A** of order n, let $|\mathbf{A}_{ij}|$ denote the determinant of the submatrix of order $(n-1)$ derived from **A** by striking out the ith row and the jth column.

Theorem A.12. $$|\mathbf{A}| = \sum_{j=1}^{n} a_{ij}(-1)^{i+j}|\mathbf{A}_{ij}|$$

No proof is offered. But let us explore an example. Let

$$\mathbf{A} = \begin{bmatrix} a_{11} & a_{12} & a_{13} \\ a_{21} & a_{22} & a_{23} \\ a_{31} & a_{32} & a_{33} \end{bmatrix}$$

and let us use the terms in the first row, i.e. set $i = 1$ in the formula given in Theorem A.12. Applying the formula, we have

$$\begin{aligned} |\mathbf{A}| &= a_{11}(-1)^2 \begin{vmatrix} a_{22} & a_{23} \\ a_{32} & a_{33} \end{vmatrix} + a_{12}(-1)^3 \begin{vmatrix} a_{21} & a_{23} \\ a_{31} & a_{33} \end{vmatrix} \\ &\qquad + a_{13}(-1)^4 \begin{vmatrix} a_{21} & a_{22} \\ a_{31} & a_{32} \end{vmatrix} \\ &= a_{11}(a_{22}a_{33} - a_{23}a_{32}) - a_{12}(a_{21}a_{33} - a_{23}a_{31}) \\ &\qquad + a_{13}(a_{21}a_{32} - a_{22}a_{31}) \end{aligned}$$

Check for yourself that this agrees with the application to this matrix **A** of the original definition of a determinant cf. example (6).

APPENDIX B

Solutions and hints for some of the exercises

Section 1.5

1. (c) $1+x+x^2$ (d) $(x_1+1)+(x_2+2)+(x_3+3)=6+x_1+x_2+x_3$

2. (a) $\sum_{i=1}^{4} a_{0i}x_i$

4. Using the three equations to eliminate q_d and q_s leaves one equation in one variable p:

$$-a+bp=c-dp$$

i.e.
$$(b+d)p=a+c$$

Since both b and d greater than zero, then $(b+d)$ not equal to zero. Thus it is possible to divide both sides by $(b+d)$ to give

$$p=(a+c)/(b+d)$$

Since all four constants are positive, this equilibrium price must be positive.

5. 200, 100 and 150 units. For method, see section 1.2.

Section 1.9

1. (b) $x<3$ (c) $x \geqq 2$ *or* $x \leqq -2$

3. $u>w-8$

4. $a+b>5$

5. (c) No feasible region. (d) Feasible region is *unbounded* (i.e. it extends indefinitely in certain directions).

7. Feasible region is

$$3x_1-4x_2 \geqq 250$$
$$-x_1+4x_2 \geqq 450$$

General condition is $(1-a_{11})/a_{12} > a_{21}/(1-a_{22})$, provided coefficients lie between 0 and 1. This condition ensures slopes such that the lines intersect in the positive quadrant.

Section 1.12

1. (e) $\{3, 4, 5\}$ (f) $\{3, 4\}$
2. (b) Valid – all elements of A are present in both of the intersecting sets.
   (c) Valid – $\varnothing$ has no elements, and hence union is just the elements in A.
4. (a) $\{x \mid 1 \leqq x \leqq 4\}$ (b) $\{x \mid -\infty \leqq x \leqq +\infty\}$
5. (a) p is necessary for q (since a square has four sides), but p is not sufficient for q (since some four-sided figures are not square).
   (e) q and r are both necessary for p; q and r *taken together* are sufficient for p.

Section 2.3

1. (d) Operation is defined since each vector has 4 elements, and both are *column* vectors. But **a** is not greater than **b** since for second element comparison, $0 < 2$.
2. $\mathbf{p} \geqq \mathbf{0}$ where $\mathbf{p}$ and $\mathbf{0}$ are column vectors of order n. (Alternatively relationship could be written in terms of row vectors: $\mathbf{p}' \geqq \mathbf{0}$.)
4. Use vector inequality relationships.
5. Price vector is $[6 \quad 1 \quad 2]$ with units as stated.

   Quantity vector is $[15 \quad 18 \quad 20]$ with units as stated.

   Total value (in pence) $= 6 \times 15 + 1 \times 18 + 2 \times 20 = 148$

   This suggests we define multiplication of one vector into another in terms of multiplying together the corresponding elements.

Section 2.6

1. Suppose that there exist scalars λ and ϕ such that

$$\begin{bmatrix} 3 \\ 14 \end{bmatrix} = \lambda \begin{bmatrix} 3 \\ 4 \end{bmatrix} + \phi \begin{bmatrix} 1 \\ -2 \end{bmatrix}$$

Now attempt to solve these equations, to see what happens.

$$\begin{aligned} 3 &= 3\lambda + \phi \\ 14 &= 4\lambda - 2\phi \end{aligned}$$

Add twice first equation to the second:

$$20 = 10\lambda$$

Thus we can find a solution $\lambda = 2 \quad \phi = -3$. The supposition that λ and ϕ exist has not led to any difficulties, and by obtaining a solution we have justified the supposition of existence. Thus

$$\begin{bmatrix} 3 \\ 14 \end{bmatrix} = 2\begin{bmatrix} 3 \\ 4 \end{bmatrix} - 3\begin{bmatrix} 1 \\ -2 \end{bmatrix}$$

(Check this vector equation for yourself.)

5. Let the set of n-component vectors be denoted $\mathbf{b}_1, \mathbf{b}_2, \ldots, \mathbf{b}_m, \mathbf{0}$. Suppose that there exist scalars λ_i such that

$$\lambda_1 \mathbf{b}_1 + \lambda_2 \mathbf{b}_2 + \ldots + \lambda_m \mathbf{b}_m + \lambda_{m+1} \mathbf{0} = \mathbf{0}$$

Obviously one set of λ_i which satisfies this vector equation is

$$\lambda_i = 0 \qquad (i = 1, 2, \ldots m)$$
$$\lambda_{m+1} = 1$$

Since we have found a solution, the initial supposition is justified. And since the λ_i are not all zero, the set is linearly dependent. (Note that it may be said, alternatively, that the null vector is linearly dependent on every other vector or set of vectors.)

6. Apply definition of linear dependence. Consider in light of exercise 5.

Section 2.8

2. The set is a spanning set but is not a basis. (In fact, any two vectors from the set form a basis.)
4. Suppose that the set is linearly dependent. Then there exist scalars λ_i not all zero such that

$$\lambda_1 (\mathbf{a}+\mathbf{b}) + \lambda_2 (\mathbf{b}+\mathbf{c}) + \lambda_3 (\mathbf{c}+\mathbf{a}) = \mathbf{0}$$

i.e. $$(\lambda_1 + \lambda_3)\,\mathbf{a} + (\lambda_1 + \lambda_2)\,\mathbf{b} + (\lambda_2 + \lambda_3)\,\mathbf{c} = \mathbf{0}$$

Since **a**, **b**, and **c** are linearly independent, then $\lambda_1 + \lambda_3 = 0$, $\lambda_1 + \lambda_2 = 0$ and $\lambda_2 + \lambda_3 = 0$. The only solution to these equations is $\lambda_1 = \lambda_2 = \lambda_3 = 0$. But this contradicts the initial supposition which thus must be wrong. Thus vectors are linearly independent.

5. If and only if $\theta = 8$, vectors do not form a basis. To show this, explore existence of scalar coefficients, as usual.

6. Regard receipts as vectors, denoted **a**, **b** and **c** respectively, and use vector inequality relationships: (i) nothing can be said, (ii) $\mathbf{c} \geqq \mathbf{a}$ and hence C is preferable to A (provided you prefer large receipts to small receipts).

Section 3.3

1. (b) Operations not defined: **A** is of order 1×2 and **B** is of order 2×1.

4. (iii) No conditions suffice; no matter what the values of α and β,

$$a_{11} > b_{11} \text{ and } a_{21} > b_{21}$$

5. (i) Little or no meaning – you cannot usefully add together unit prices for the two years.
   (ii) No meaning whatsoever – you cannot add prices to quantities since they are entirely different dimensions.
   (iii) The element in the ith row and j^{th} column of $\mathbf{Q}+\mathbf{S}$ is the total quantity sold during the two years of the ith product in the jth shop.

Section 3.5

2. tr $\mathbf{D} \geqq 2n$.

3. $\mathbf{B}_1$ is identity matrix. $\mathbf{B}_2$ has zero in first diagonal position, unit entries on remainder of principal diagonal (i.e. beginning in second position) and zero entries elsewhere. In $\mathbf{B}_3$, unit entries on principal diagonal begin in third position, and so forth. In the matrix comprising the sum of all these matrices, all elements off diagonal are zero. On the diagonal, first element is 1 (from $\mathbf{B}_1$); second element is 2 (comprising 1 from $\mathbf{B}_1$ and 1 from $\mathbf{B}_2$); third entry is 3, and so forth. But this is just the same as **A**.

Section 3.7

1. $\mathbf{a}'\mathbf{b}$ is defined and equals 7

2. (a) $\mathbf{AB} = \begin{bmatrix} 4 & -2 & 1 \\ 4 & 0 & 2 \end{bmatrix}$

   (c) **A** has 2 columns, **C** has 3 rows. Thus **AC** is not defined.

Section 3.9

1. (c) **A** has 4 columns, **B** has 3 rows. Thus **AB** is not defined. But **BA** is defined:

$$\mathbf{BA} = \begin{bmatrix} 6 & 21 & 12 & -3 \\ 14 & 49 & 28 & -7 \\ 4 & 14 & 8 & -2 \end{bmatrix}$$

2. A set of necessary and sufficient conditions is (i) $\alpha = -2$ *and* (ii) $\beta = -2$.

Section 3.12

1. (b) $a_{i,\, j+1}$ is element in i^{th} row and $(j+1)^{\text{th}}$ column.

$$\sum_{i=1}^{2} (a_{i1}\, a_{i2} + a_{i2}\, a_{i3}) = (3\times 4 + 4\times(-1)) + (0\times 2 + 2\times 1)$$
$$= 10$$

2. (a) Total number of units of ith commodity held at all branches considered together.
   (b) Little or no meaning – it relates to j^{th} branch, but you cannot add up units of different commodities (unless you simply want to know the number of objects – of whatever kind – at the jth branch).
4. $\mathbf{CBA\,x} = \mathbf{CBg} - \mathbf{h}$, a system of p equations.

Section 3.15

2. There are several alternative ways of partitioning **A**; a computationally efficient approach is to partition as follows to exploit the structure of **A**:

$$\mathbf{A} = \left[\begin{array}{cc|cc} 3 & 4 & 0 & 0 \\ 2 & 7 & 0 & 0 \\ \hline -2 & 0 & 1 & 0 \\ 0 & -2 & 0 & 1 \end{array}\right] = \begin{bmatrix} \mathbf{A}_{11} & \mathbf{0} \\ -2\mathbf{I} & \mathbf{I} \end{bmatrix}$$

To permit partitioned multiplication, **B** must then be partitioned into two square submatrices. Check your partitioned multiplication by directly multiplying **A** into **B**.

4. Note that if **A** is square, $\mathbf{A}+\mathbf{A}'$ is defined,

$$\begin{aligned} (\mathbf{A}+\mathbf{A}')' &= \mathbf{A}' + (\mathbf{A}')' && \text{by Theorem 3.1} \\ &= \mathbf{A}' + \mathbf{A} \\ &= \mathbf{A} + \mathbf{A}' \end{aligned}$$

Thus $\mathbf{A}+\mathbf{A}'$ is symmetric. (Analogous proof for $\mathbf{A}-\mathbf{A}'$.)

6. Using the obvious notation, model is $\mathbf{x}=\mathbf{Ax}+\mathbf{b}$. This may be written

$$\mathbf{Ix}=\mathbf{Ax}+\mathbf{b}$$

i.e.

$$(\mathbf{I}-\mathbf{A})\mathbf{x}=\mathbf{b}$$

7. Let q_{ik} denote weight (in appropriate units) of ith ingredient to be delivered to k^{th} bakery ($i=1,2,3;\ k=1,2$). Then

$$\mathbf{Q}=\mathbf{AX}=\begin{bmatrix}157 & 498\\ 35 & 75\\ 29{\cdot}5 & 54{\cdot}5\end{bmatrix}$$

12. Since $\mathbf{A}$ is symmetric, it must be square. Let m denote its order. Then, to permit multiplication, $\mathbf{B}$ must have m rows. Let number of columns in $\mathbf{B}$ be denoted n. Then $\mathbf{B}'\mathbf{AB}$ is square, of order n.

$$\begin{aligned}(\mathbf{B}'\mathbf{AB})' &= \mathbf{B}'\,\mathbf{A}'\,(\mathbf{B}')' && \text{by Theorem 3.3}\\ &= \mathbf{B}'\,\mathbf{A}'\,\mathbf{B}\\ &= \mathbf{B}'\,(-\mathbf{A})\,\mathbf{B} && \text{since } \mathbf{A} \text{ is skew-symmetric}\\ &= -\mathbf{B}'\,\mathbf{A}\,\mathbf{B}\end{aligned}$$

Thus $\mathbf{B}'\,\mathbf{AB}$ is skew-symmetric.

14. $\mathbf{A}+\mathbf{B}\geqq\mathbf{C}+\mathbf{D}$ provided (i) $\alpha\leqq 0$ *and* (ii) $\beta=1$.

Section 4.3

3. The required premultiplying matrix is

$$\mathbf{E}=\begin{bmatrix}1 & 0 & -1\\ 0 & 1 & -2\\ 0 & 0 & 1\end{bmatrix}$$

Then

$$\mathbf{EA}=\begin{bmatrix}-1 & -4 & 3\\ -3 & -6 & 5\\ 2 & 4 & -1\end{bmatrix}$$

Section 4.5

1. *An* echelon matrix (obtained without interchanging any rows) is

$$\begin{bmatrix}1 & \frac{3}{2} & 3 & \frac{5}{2}\\ 0 & 1 & -\frac{4}{3} & 1\\ 0 & 0 & 1 & -\frac{3}{17}\end{bmatrix}$$

Section 4.7

1. $\begin{bmatrix} 1 & 2 \\ 0 & 1 \\ 0 & 0 \end{bmatrix} \qquad 3\mathbf{r}_1 - \mathbf{r}_2 - \mathbf{r}_3 = \mathbf{0}$

Section 4.11

2. By way of an aside, note that in **BA**, $\mathbf{r}_1 - 2\mathbf{r}_2 + \mathbf{r}_3 = \mathbf{0}$ (where $\mathbf{r}_i$ denote rows of **BA**) and that *same* relationship holds for rows of **B**. (This follows from the definition of matrix multiplication – check for yourself.)
3. $r(\mathbf{A}) = 3$, i.e. **A** is non-singular. Thus from Theorem 4.4, $r(\mathbf{AB}) = 1$ if and only if $r(\mathbf{B}) = 1$. Any echelon matrix derived from **B** must have two rows of zero entries if $r(\mathbf{B})$ is to be made equal to one. When an echelon is computed, we find this can be achieved if $\alpha = \frac{3}{2}$ *and* $\beta = 4$. (Alternatively we can find an echelon matrix from **AB** and then obtain these conditions on α and β. But such an approach requires more algebraic manipulation.)
5. Rank of **C** is 1.

Section 5.3

3. Procedure breaks down because echelon (shown below) does not have unit entries in all the positions in the principal diagonal.

$$\begin{bmatrix} 1 & \frac{1}{2} & 1 \\ 0 & 1 & 2 \\ 0 & 0 & 0 \end{bmatrix}$$

Section 5.7

2. Inverse exists if diagonal matrix is non-singular; this requires all $\lambda_i \neq 0$ (cf. exercise 4 of section 4.11). To transform **D** to an identity matrix, multiply i^{th} row by λ_i^{-1}. Then the inverse matrix is the product of the elementary matrices corresponding to these multiplications. Hence provided all $\lambda_i \neq 0$,

$$\mathbf{D}^{-1} = \begin{bmatrix} \lambda_1^{-1} & & & & 0 \\ & \lambda_2^{-1} & & & \\ & & \cdot & & \\ & & & \cdot & \\ & 0 & & & \cdot \\ & & & & \lambda_n^{-1} \end{bmatrix}$$

Section 5.9

1. Observe that, if the initial partitioned matrix is denoted $\begin{bmatrix} \mathbf{E} & \mathbf{F} \\ \mathbf{G} & \mathbf{H} \end{bmatrix}$ where $\mathbf{F} = \mathbf{0}$ and $\mathbf{H} = \mathbf{I}_2$, then the inverse is $\begin{bmatrix} \mathbf{E}^{-1} & \mathbf{0} \\ \mathbf{GE}^{-1} & \mathbf{I}_2 \end{bmatrix}$ as may be shown by applying the usual rules for partitioned multiplication.

2. The partitioning approach breaks down. But note that this does *not* prove that the initial matrix is singular.

3.
$$\mathbf{B}^{-1} = \tfrac{1}{19} \begin{bmatrix} 1 & 2 & -8 & 11 \\ -10 & -1 & 23 & -15 \\ 4 & 8 & -13 & 6 \\ 7 & -5 & 1 & 1 \end{bmatrix}$$

4. Since $\mathbf{A}$ is non-singular, each $\mathbf{A}_j$ must be non-singular. To find $\mathbf{A}^{-1}$, apply elementary row operations to reduce $\mathbf{A}$ to an identity matrix. When these operations are applied to rows corresponding to $\mathbf{A}_1$, associated matrix will have $\mathbf{A}_1^{-1}$ in corresponding rows. Similarly for all other $\mathbf{A}_j$. Thus

$$\mathbf{A}^{-1} = \begin{bmatrix} \mathbf{A}_1^{-1} & & & & & \\ & \mathbf{A}_2^{-1} & & & 0 & \\ & & \cdot & & & \\ & & & \cdot & & \\ & 0 & & & \cdot & \\ & & & & & \cdot \\ & & & & & \mathbf{A}_n^{-1} \end{bmatrix}$$

(As always, the result may be checked by computing $\mathbf{AA}^{-1}$.)

Section 6.4

1. The coefficients matrix is successfully reduced to an identity matrix, and hence it is non-singular. The solution is $[x_1 \quad x_2 \quad x_3] = [1, 2, 3]$ and by Theorem 6.1 it is unique.

2. Note that there is a *unique* solution, by the same argument as for exercise 1.

Section 6.6

1. No solution, by Theorem 6.2. (Note that equation four is inconsistent with the first equation.)

2. $[x_1 \quad x_2 \quad x_3] = [1 \quad 2 \quad 3]$.

Section 6.8

1. $r(\mathbf{A}) = r(\mathbf{U})$ and hence solution exists (by Theorem 6.3). Since $k = 2$ and $n = 3$, *one* variable can be assigned a parametric value. From transformed matrix, general solution is

$$\begin{bmatrix} x_1 \\ x_2 \\ x_3 \end{bmatrix} = \begin{bmatrix} 5-1{\cdot}5\,\theta \\ 2-0{\cdot}5\,\theta \\ \theta \end{bmatrix}$$

5. $r(\mathbf{A}) = r(\mathbf{U})$ and hence solution exists. Since $k = 2$ and $n = 3$, one variable can be assigned a parametric value. From transformed matrix, general solution is

$$\begin{bmatrix} x_1 \\ x_2 \\ x_3 \end{bmatrix} = \begin{bmatrix} 1 \\ -2+\theta \\ \theta \end{bmatrix}$$

Note that the value for x_1 is unique (and does not depend on θ). In general, when some 'spare' variables occur and we assign parametric values, each of the remaining variables may be a function of all the parameters, some of the parameters, or none of the parameters. (Consider also the solution of the system (6–11) in section 6.7.) Note also that the second and third columns of $\mathbf{A}$ are linearly dependent; column 3 is (-1) times column 2. If the variables were listed in the order x_2, x_3, x_1, this linear dependence would prevent us obtaining an identity matrix in the first two columns of the transformed matrix. This is just as well, because otherwise we could set $x_1 = \phi$ (a parameter) and this would contradict the previous result that there is the unique value $x_1 = 1$.

Section 6.10

4. $m = 2$, $n = 3$ and $k = 2$. Thus infinite numbers of solutions exist. General solution is $[x_1 \quad x_2 \quad x_3] = [0 \quad \theta \quad \theta]$. Note that the value for x_1 is unique. When this happens (in the context of homogeneous equations), the unique value must be zero, since the

trivial solution always exists. Note that the columns corresponding to the other two variables are linearly dependent on each other.

5. Note that the two solutions are *not* multiples one of the other. But multiples of either of these two solutions are also solutions.

Section 6.13

3. From the viewpoint of mathematical analysis, the general solution is

$$\begin{bmatrix} x_1 \\ x_2 \\ x_3 \end{bmatrix} = \begin{bmatrix} 5-\theta \\ 2-\theta \\ \theta \end{bmatrix}$$

But in the applied context, negative purchases are not meaningful. Thus we must add the qualification 'for all values of θ in the range $0 \leqq \theta \leqq 2$' in order to ensure that each element in the solution vector is non-negative.

4. Let the tonnages of the three commodities be denoted x_1, x_2 and x_3. Then gross revenue (in £) is

$$4x_1 + 8x_2 + 2x_3 = 100$$

and the wholesaler's outlay (in £) is

$$3{\cdot}5x_1 + 7x_2 + 1{\cdot}5x_3 = 85{\cdot}5$$

These two equations in three variables may be solved in the usual way to give $[x_1 \quad x_2 \quad x_3] = [21-2\theta \quad \theta \quad 8]$. This shows immediately that he bought 8 tons of commodity C. For the other two commodities, we can give only a range of values:

quantity of A is $(21-2\theta)$ tons
quantity of B is θ tons

where θ is a parameter in the range $0 \leqq \theta \leqq 21/2$ (to ensure that quantities purchased are non-negative). In mathematical terms, we get a unique answer for x_3 (quantity of C) because the *other* two columns are linearly dependent, and hence it is not possible to assign a parametric value to x_3. One way of interpreting this in the applied context is to observe that he spends £12 on commodity C, but that the only other fact we can infer is that he spends a total of £73·5 on commodities A and B together. We are unable to separate these two commodities because the ratio of profit to selling price is the same (i.e. the corresponding columns in the coefficients matrix are linearly dependent), and hence no matter how his expenditure of £73·5 is distributed between the two commodities, the total profit and the total gross revenue (for these two commodities taken together) is the same, at £10·5 and £84 respectively. We simply do not have enough information to enable us to find out precisely how he divided his total expenditure on these two commodities; but we do know that the division must correspond to some particular value for θ in the above expressions.

Section 6.15

1. Assume $c_{11} \neq 0$ and $c_{11}c_{22} \neq c_{12}c_{21}$.
   Then $p_2 = (c_{01}c_{12} - c_{02}c_{11})/(c_{11}c_{22} - c_{12}c_{21})$.

2. Let the numbers in the four categories be denoted

yes/yes	x_1
yes/no	x_2
no/yes	x_3
no/no	x_4

From the counts already made,

$$\begin{aligned} x_1 + x_2 \qquad\qquad &= 1738 \\ x_1 \qquad + x_3 \qquad &= 1675 \\ x_3 + x_4 &= 1402 \\ x_2 \qquad + x_4 &= 1465 \end{aligned}$$

The problem now is: can these equations be solved? The standard technique may be employed – form the augmented matrix and carry out elementary row operations until the following position is reached:

$$\begin{matrix} 1 & 1 & 0 & 0 & 1738 \\ 0 & 1 & -1 & 0 & 63 \\ 0 & 0 & 1 & 1 & 1402 \\ 0 & 0 & 0 & 0 & 0 \end{matrix}$$

Note that $3 = r(\mathbf{A}) = r(\mathbf{U}) < n = 4$.

Thus the equations have an infinite number of solutions (as they stand), with one arbitrary parameter (see case III of Table 6.11). In other words, one more piece of information is required before we can establish which of the range of solutions to this set of equations is the one which is relevant. To meet this need, we could directly measure x_4 (by sorting the questionnaires into two piles – those answering 'no/no', and all others). Once the value of x_4 is known, the values of the other three variables can be found by back-substitution, using the above tableau.

3. $\mathbf{A}+\mathbf{B} = \begin{bmatrix} 3{\cdot}5 & 4 \\ 4 & \alpha+\beta \end{bmatrix} \qquad \mathbf{AB} = \begin{bmatrix} 1{\cdot}5 & 6+2\beta \\ 2 & 8+\alpha\beta \end{bmatrix}$

(i) A set of necessary and sufficient conditions for $\mathbf{A}+\mathbf{B} > \mathbf{AB}$ is

$$4 > 6+2\beta \qquad \text{i.e. } \beta < -1 \qquad \text{(I)}$$

and

$$\alpha+\beta > 8+\alpha\beta \qquad \text{i.e. } \alpha(1-\beta) > 8-\beta \qquad \text{(II)}$$

(ii) If we take the necessary condition $\beta < -1$ as given, then

$$1-\beta > 2 \qquad \text{(III)}$$

Since in this case $(1-\beta)$ is strictly positive, we may divide both sides of (II) above without changing the direction of the inequality, to give

$$\alpha > \frac{8-\beta}{1-\beta}$$

i.e.

$$\alpha > 1+\frac{7}{1-\beta}$$

From (III)

$$\frac{1}{1-\beta} < \tfrac{1}{2}$$

and thus

$$1+\frac{7}{1-\beta} < 1+\tfrac{7}{2}$$

Thus $\alpha > \frac{9}{2}$ is sufficient to ensure $\alpha > \dfrac{8-\beta}{1-\beta}$

In summary, a set of conditions which is sufficient (but not necessary) to ensure that $\mathbf{A}+\mathbf{B} > \mathbf{AB}$ is

(a) $\alpha > \frac{9}{2}$

and (b) $\beta < -1$

4. Let the numbers of acres devoted to barley and to oats be denoted x_1 and x_2 respectively. Clearly we need $x_1 \geqq 0$ and $x_2 \geqq 0$ as well as the following constraints:

(a) land used $\leqq$ land available

i.e. $$x_1 + x_2 \leqq 100$$

(b) labour used $\leqq$ labour available

Each acre in barley requires $5 + (0{\cdot}2 \times 50) = 15$ man-hours. Similarly each acre in oats requires 10 man-hours. Thus the constraint is

$$15x_1 + 10x_2 \leqq 1200$$

(It is not valid to suppose that all 100 acres will be cultivated; and hence we should *not* assign 500 man-hours immediately on a per-acre basis, and then add per-bushel requirements.)

(c) transport capacity used $\leqq$ transport capacity available

i.e. $$50x_1 + 80x_2 \leqq 6400$$

Finally, the gross receipts (in £) to be maximized is

$$50 \times 0{\cdot}8x_1 + 80 \times 0{\cdot}9x_2 = 40x_1 + 72x_2$$

The feasible region can be graphed as in section 1.8. (An alternative algebraic formulation is to define two variables measuring the number of bushels of oats and barley.)

5. Let the vector of initial staff numbers be $\mathbf{x}_0$, and those at the end of years 1 and 2 be $\mathbf{x}_1$ and $\mathbf{x}_2$ respectively. Let the annual recruitment numbers in the three grades be denoted by the vector $\mathbf{w}$. Suppose, for the present, that there is a feasible solution for $\mathbf{w}$. Now

$$\mathbf{x}_1 = \mathbf{P}' \mathbf{x}_0 + \mathbf{w}$$

and $$\mathbf{x}_2 = \mathbf{P}' \mathbf{x}_1 + \mathbf{w}$$

Substitute in the second equation for $\mathbf{x}_1$ to give

$$\begin{aligned}\mathbf{x}_2 &= \mathbf{P}'(\mathbf{P}' \mathbf{x}_0 + \mathbf{w}) + \mathbf{w} \\ &= (\mathbf{P}')^2 \mathbf{x}_0 + \mathbf{P}'\mathbf{w} + \mathbf{w} \\ &= (\mathbf{P}')^2 \mathbf{x}_0 + (\mathbf{P}' + \mathbf{I}_3)\mathbf{w}\end{aligned}$$

Thus $$\mathbf{w} = (\mathbf{P}' + \mathbf{I}_3)^{-1}\{\mathbf{x}_2 - (\mathbf{P}')^2 \mathbf{x}_0\}$$

provided the inverse matrix exists. This shows how to calculate **w**, all of whose elements are required to be non-negative. Use elementary row operations in the usual way to invert $(\mathbf{P}'+\mathbf{I}_3)$; it turns out that the inverse does exist, and is

$$\begin{bmatrix} 0{\cdot}625 & 0 & 0 \\ -0{\cdot}1014 & 0{\cdot}5405 & 0 \\ 0{\cdot}0053 & -0{\cdot}0284 & 0{\cdot}5263 \end{bmatrix}$$

The vector **w** is then found to be [200 204 15] after rounding each element to the nearest whole number. Since each element is non-negative, the employer *can* achieve all his aims.

Section 7.3

1. There are five basic solutions, of which three are degenerate. Note that in augmented matrix, column 2 is twice column 1, and last column is three times the fourth column.
2. The non-negative basic solutions correspond to the 'corners' of the feasible region.
3. In the augmented matrix, the sum of the first two columns is the same as the last column.

Section 7.5

1. There are four basic feasible solutions. That which gives the highest profit (of 50 money units) is [10 20 0 0].
2. Function to be maximized is $-2x_1+x_2$ (since maximum of (-1) times a function = minimum of the function; consider a graph of a function such as $y=x^2$). Second constraint may be multiplied by (-1); this changes direction of inequality (cf. section 1.6) and hence constraint becomes $-2x_1-x_2 \leqq 1$.

Section 7.7

1. (a) $x_1{}^2-4x_1x_2+5x_2{}^2$

2. (a) $\mathbf{A} = \begin{bmatrix} 1 & 1 \\ 1 & -2 \end{bmatrix}$

3. (i) $(x_1-2x_2)^2+x_2{}^2$ positive definite
   (iv) $(x_1-x_2)^2$ positive semi-definite

Section 7.10

2. (a) $(t, 2t)$ (b) $(1+t, 2-t)$ or any equivalent expression
   (c) No solution

3. Using two distinct arbitrary integers s and t, the general solutions to separate equations are $(s, 2s)$ and $(1+t, 2-t)$. Problem now is: can we find particular *integer* values for s and t which make these solutions identical. Thus we want to solve pair of equations

$$s = 1+t \quad \text{and} \quad 2s = 2-t$$

   These have *unique* solution $s = 1$, $t = 0$ and it happens that both are integers. These correspond to a solution for each of the two original equations of (1, 2). In other words, this solution is common to both sets of solutions for the separate equations. Hence system has a unique solution (1, 2). Graphically, for each separate equation, the locus of solutions is a set of equally spaced points lying along a straight line. The two lines intersect, and it happens that the point of intersection (1, 2) is a member of both sets of solutions. An alternative algebraic approach is simply to solve the equations in continuous variables and see if solution is in integers.

4. No integer solution.

Section 7.12

1. (a) General solution in continuous variables is $[(1+\theta/3)\ (1—\theta/3)\ \theta]$. For non-negative integer solution, set $\theta = 0$ or 3.

2. Hint: which strictly positive integers are small enough to satisfy the second equation?

3. Andrews is elected with 3 votes at first count and 5 at second.

Section 7.15

1. Two feasible basic solutions, both entirely in integers.

2. Optimal solution is: from first depot, 3 cars to first garage, 2 to second; all cars from second depot to second garage. Note that to move a car from second depot to second garage is the most expensive of the individual unit costs. Why is this move used in the optimal solution?

APPENDIX C

Suggestions for further reading

This appendix lists some books which the reader may consult in order to follow up the references made in various chapters to further developments, or to obtain a deeper knowledge of some of the ground covered in the text. References are made only to mathematical techniques; no attempt has been made to list work on the various social science problems which have been introduced as illustrations.

Chapter 1. A deeper and more fundamental discussion of the elements of set theory is to be found in Chapter 4 of R. G. D. Allen, *Basic Mathematics* (Macmillan, London, 1964). Necessary and sufficient conditions are similarly treated in section 5.3 of the same book.

Chapter 4 (with other chapters). The main theme of linear dependence of vectors and the classification of matrices according to their rank can be treated in a slightly deeper fashion using further concepts (e.g. dimension) relating to vector spaces. And such a treatment can be used (*inter alia*) to give a proof of Theorem 4.2 (or of an equivalent result). Among many books which pursue this path, J. R. Munkres, *Elementary Linear Algebra* (Addison-Wesley, Reading, Massachusetts, 1964) is of interest. Although the book is terse, anyone who has mastered the present text should be able to follow the proofs which relate to this theme; unlike many such books, determinants are not used in the course of developing the theme. Another exposition, which does however use determinants, is to be found in Chapter 13 of R. G. D. Allen's book.

Chapter 5. Some further details on practical computation and rounding errors in matrix inversion can be found in sections 1.3 and 1.4 of Lucy Joan Slater, *Fortran Programs for Economists* (Cambridge University Press, London, 1967).

Chapter 7. There are a very large number of text-books on linear programming. And the mathematical theory of linear programming is substantial. A systematic, detailed treatment of the ground covered in section 7.4, and of one algorithm for computing solutions, is to be found in Chapters 3 and 4 of G. Hadley, *Linear Programming* (Addison-Wesley, Reading, Massachusetts, 1962); and sections 17 to 22 of Chapter 2 of that book deal with point sets and convexity, which are mathematical tools required for linear programming theory but not covered in the present text. A short account of similar aspects of linear programming is to be found in Chapters IV to VI of S. Vajda, *The Theory of Games and Linear Programming* (Methuen, London, 1956).

One computationally efficient approach for diagonalizing a quadratic form depends on certain further matrix properties, notably latent roots and eigenvectors; all these matters may be studied in Chapter 7 of G. Hadley, *Linear Algebra* (Addison-Wesley, Reading, Massachusetts, 1961).

Further study of a single linear equation in two integer variables is perhaps mainly of interest to pure mathematicians; however, the systematic method of solution (mentioned at the end of section 7.9) may be found in section 2.4 of W. J. LeVeque, *Topics in Number Theory* (Vol. 1) (Addison-Wesley, Reading, Massachusetts, 1956), and in section 2 of A. O. Gelfond, *The Solutions of Equations in Integers* (W. H. Freeman and Co., San Francisco, 1961).

The test on the **A** matrix (mentioned at the end of section 7.14) to check whether a linear programming problem is bound to have an integer solution, is in effect a check as to whether **A** is 'unimodular'; this term is explained in section 9.2 of Hadley's book on linear programming. A more detailed statement of useful sufficient conditions for unimodularity is given in Theorems 3, 5 and 6 in Chapter 13 of H. W. Kuhn and A. W. Tucker (Editors), *Linear Inequalities and Related Systems* (Princeton University Press, Princeton, 1956). (This is an advanced mathematical work, but the statements of these theorems are not difficult to follow.)

Index